Fundamentals and Applications of
Digital Logic Circuits

Sol Libes
Union County Technical Institute

HAYDEN BOOK COMPANY, INC.
Rochelle Park, New Jersey

Library of Congress Cataloging in Publication Data

Libes, Sol.
    Fundamentals and applications of digital logic
    circuits.

    1. Digital electronics. 2. Logic circuits.
I. Title.
TK7868.D5L52   621.3815'3   74-16173
ISBN 0-8104-5506-4
ISBN 0-8104-5505-6 (pbk.)

Printed in the United States of America

          3   4   5   6   7   8   9   PRINTING

        76 77 78 79 80 81 82   YEAR

Preface

Integrated circuits and digital logic have, over the past decade, led to the development of electronic equipment of increasing sophistication and complexity. Where previously a service technician could expect to encounter tens or at most a few hundred devices in an electronic system, he may now have thousands and even millions of components. System reliability and maintainability have increased. However, the technician who undertakes to maintain this equipment must have a good grasp of the inner workings of these complex electronic systems.

This book is intended as an introduction to digital logic circuits and many of their applications, especially the digital instruments used in industry, such as digital voltmeters. A brief discussion of computers is limited primarily to the electronic circuitry employed. The emphasis, throughout, is on semiconductor integrated circuits, inasmuch as this type of electronic circuitry is in popular use today. Therefore, the applications of functional rather than discrete semiconductor theory are stressed.

The author wishes to express his appreciation to the following individuals and companies for their assistance in furnishing information and granting permission to use material: Ed Lord, Burroughs Corp.; Carol Parker, Fairchild Semiconductor; Robert Sumbs, Intel Corp; and Ed Klebaur, Weston Instruments.

SOL LIBES

Springfield, New Jersey

Contents

Chapter 1

Semiconductor Fundamentals

The theory of semiconductor devices (diodes and transistors) is covered in any number of excellent textbooks. Nevertheless, a brief review of their characteristics as *switching devices* deserves some examination before delving into digital logic circuits.

CONDUCTORS, NONCONDUCTORS, AND SEMICONDUCTORS

A conductor is a material having little opposition, and hence little resistance, to the flow of an electric current through it. Copper is a good conductor since it has a very low resistance. For example, one foot of No. 18 AWG copper wire (widely used in wiring electronic equipment), which has a diameter of approximately 0.04 inch and has a resistance of 0.0065 ohm.

A nonconductor has a very high resistance and therefore offers a very great amount of opposition to the flow of electric current. A piece of polystyrene the same size as No. 18 AWG copper wire would have a resistance greater than 10^{15} ohm. As virtually no current will flow through a good nonconductor, it can be used to insulate conductors.

A semiconductor's resistance lies between that of a conductor and a nonconductor. A typical semiconductor has a resistance of approximately 10 ohms per cubic centimeter. Silicon and germanium are the semiconductor materials most often used in such semiconductor devices as diodes and transistors.

THE DIODE

The standard diode is constructed by joining opposite types (N and P types) of semiconductor material. The N material has been doped with an impurity that creates surplus electrons which serve as the current carriers. The P material has been doped with an impurity that creates a deficiency of electrons, or "holes," which serve as current carriers. The electron carriers have a negative charge, while the holes have a positive charge, as shown in Fig. 1-1. The opposite charges are attracted to the junction, but cannot cross over because of the

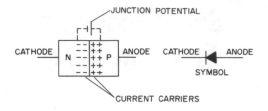

Fig. 1-1. Construction of a diode.

physical barrier of the junction. Thus, a potential difference exists at the junction. In the case of germanium, the potential is approximately 0.3 volt. Silicon has a potential of approximately 0.6 volt. This junction-potential is shown in Fig. 1-1.

When a potential of the polarity shown in Fig. 1-2, is applied to the diode, the carriers move closer to the junction. When the applied voltage exceeds the junction potential, the carriers cross over the junction and current* flows through the diode. We say that the diode is "turned on" or "forward-biased."

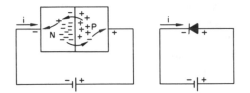

Fig. 1-2. The forward-biased diode.

When the battery is reversed, the carriers are repelled from the junction. No carriers cross the junction and hence no current flows. The diode is "turned off," and is said to be "reverse-biased." If there are impurities in the semiconductor material, a small reverse current will flow. This current is referred to as the "leakage current."

Since a high current flows in the forward-biased state, and virtually no current flows in the reverse-biased state, the diode acts as a switch. The exceptions being that: There is a forward voltage drop equal to the junction potential (0.3 volt for germanium and 0.6 volt for silicon), while a switch has no forward voltage drop. There is a small reverse leakage current, while a switch has no reverse current.

We can therefore say that the diode has a low resistance in the forward direction and a very high resistance in the reverse direction. This forward-to-

*This text uses the direction of electron flow as the direction of current flow.

reverse resistance ratio can be measured with an ohmmeter, and in fact is a popular method used when checking diodes.

Diodes are frequently employed to construct logic gates. A typical diode logic gate circuit is shown in Fig. 3-10.

THE JUNCTION TRANSISTOR

A junction-type transistor is constructed by creating a reverse-biased junction, as shown in Fig. 1-3, referred to as the *base* and the *collector*. Therefore, only a tiny reverse leakage current flows. Now, if a second junction, shown in Fig. 1-4A, referred to as the *emitter-to-base* junction, is formed and

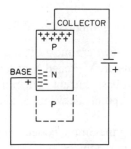

Fig. 1-3. The reverse-biased base-to-collector junction of a transistor.

forward-biased, additional carriers will be injected into the base area from the emitter. These carriers cross over into the collector creating a current flow from emitter to collector. Hence, only a small forward current flow between emitter and base will cause a large current flow between emitter and collector. The transistor thus amplifies the small base current into the large collector current. When there is no base current into the large collector current. A small amount of base current can cause maximum collector current to flow. The ratio of

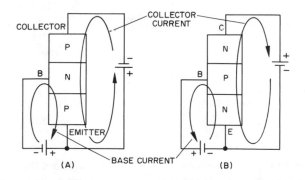

Fig. 1-4. (A) Biasing of a PNP transistor; (B) the NPN transistor.

base-to-collector current change is the measure of the transistor's gain, and is referred to as *beta* (Greek symbol β).

The transistor shown in Fig. 1-4A is a PNP type transistor. In a similar manner, an NPN transistor is constructed as shown in Fig. 1-4B. The currents flow in the opposite direction to that of the PNP, and therefore the potentials are reversed. The schematic symbols for each are shown in Fig. 1-5.

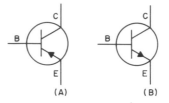

Fig. 1-5. (A) PNP symbol; (B) NPN symbol.

When transistors are operated in logic circuits they are most often used as electrical switches and hence their collector currents are switched between zero and maximum by applying either zero or maximum base current. When transistors are used as signal amplifiers they are operated at points between the zero and maximum limits, but do not reach these limits.

The transistor is said to be *on* when it is passing maximum collector current and is *off* when not passing current. The transistor may thus be switched on and off. Such a circuit is shown in Fig. 1-6.

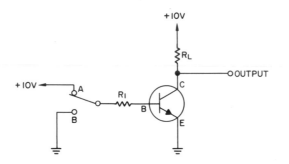

Fig. 1-6. A transistor switching circuit.

Referring to Fig. 1-6, with the switch in the *A* position, the base-to-emitter junction is forward-biased causing the collector current to be maximum. Since the collector current is maximum, the collector-to-emitter resistance is very low and the voltage at the output will be close to ground potential, the voltage being dropped across R_L. When the switch is moved to position *B*, the base-to-emitter junction is no longer forward-biased, causing the emitter-to-collector current to be zero. Since the collector current is zero, the collector-to-emitter resistance is maximum, causing the voltage at the output to equal the supply voltage of 10 volts (no voltage is dropped across R_L). The output will thus switch from essentially 0 to 10 volts, as the input switches from 10 to 0 volts.

When you consider that more than one input (refer to Figs. 3-1 and 3-3) or switching device, such as photocells, can replace the switch in Fig. 1-6, you can see that a transistor is a very useful switching device.

TRANSISTOR TYPES

Since the invention of the transistor in the 1940s, its construction, performance, and reliability have changed rapidly. The earliest devices were *point-contact* types, in which two pointed wires contacted a piece of semiconductor material to form P-N junctions. Because of its limited capabilities, this type has long been superseded by alloy, mesa, and epitaxial type transistors.

Alloy Transistors

This transistor (Fig. 1-7A) is formed from a small substrate section (1.8-inch square × 0.01-inch thick) of a semiconductor crystal. Depending on the type of crystal, two small dots a P-type material are melted against opposite surfaces of the substrate to form a PNP device, or two small dots of N-type to form an NPN device. Such a device has a high current gain but is limited in frequency response because of the wideness of the base.

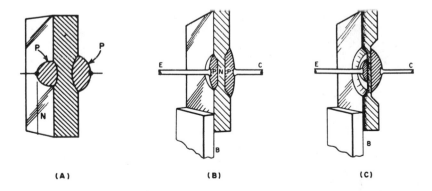

(A)   **(B)**   **(C)**

Fig. 1-7. Construction details of alloy-type transistors; (A) standard alloy type, (B) micro-alloy MAT type, and (C) alloy-diffused MADT type.

The frequency-handling capabilities are improved by etching pits into each surface of the substrate (Fig. 1-7B) to create an extremely thin base region. This structure is called the microalloy transistor (MAT). The improvement in frequency response, however, is gained at the expense of voltage-handling ability, which is limited by the narrow base region.

A further improvement of the alloy-type transistor is achieved by diffusing impurities into the base region to form a nonuniform concentration of impurities. This procedure reduces the gain of the transistor somewhat but extends its switching speed to 100 microseconds. A transistor of this type (Fig. 1-7C) is commonly referred to as a *drift* transistor or MADT (microalloy-diffused transistor).

Mesa Transistors

A major breakthrough in transistor design came with the development of the *mesa* structure. Using photolithographic and masking techniques for etching and diffusion, increased control was gained over junction spacing and impurity levels, thereby improving performance and reducing manufacturing costs at the same time.

The grown crystal is in the form of a rod *sliced* into very thin wafers (approximately 1 inch in diameter) that will serve as collectors for thousands of transistors (Fig. 1-8A). An opposite polarity impurity is then diffused into the wafer to form a two layer PN (or NP) junction. The diffusion, which will serve

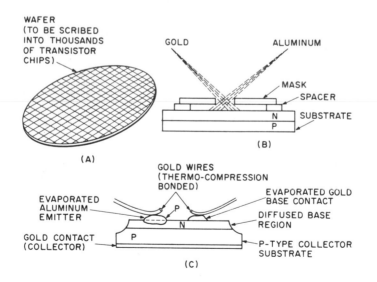

Fig. 1-8. Fabrication of diffused-case mesa transistor; (A) the wafer, (B) cross evaporation of emitter region and base contact, and (C) cross section.

as the base region, is approximately 0.0001 inch in depth. The emitter junction is formed by the evaporation or diffusion of an impurity (the same as that of the collector) through a metal mask into the base region (Fig. 1-8B).

The masking process is such that thousands of emitters are formed simultaneously and uniformly. The area between transistors is etched to form a plateau (mesa), and the wafer is now scribed and broken to obtain the individual chips. Mesa transistors can work at higher voltages than alloy types: they are stronger, have very high gain, and can work at switching speeds as fast as 1 micro-microsecond.

Epitaxial Transistors

For the epitaxial transistor, a collector crystalline film of any desired impurity concentration and thickness is grown on top of a single crystal substrate, as shown in Fig. 1-9, A and B. On top of this is grown a very thin film of silicon dioxide (SiO_2), an insulating material that protects the epitaxial layer against penetration of impurities both during and after diffusion cycles. Using a

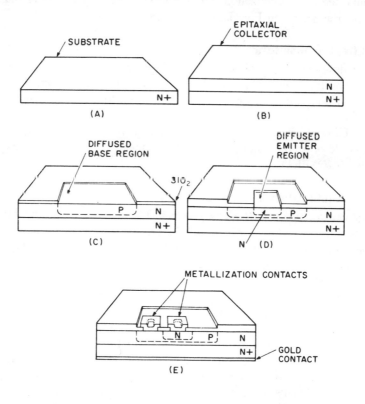

Fig. 1-9. Fabrication of an epitaxial-type transistor.

masked photo-resist process, the SiO_2 is then removed from those areas where the base diffusion is to take place (Fig. 1-9C). The mask used for the photo-resist contains multiple apertures so that thousands of bases may be formed during a single diffusion process.

The base regions are diffused into the collector's epitaxial layer. Another layer of SiO_2 is grown over the entire wafer to "passivate" the base-emitter junction, that is, to protect it against impurity penetration.

A second photo-resist diffusion and passivation process is now used to form the emitter regions (Fig. 1-9D). Openings are etched in the SiO_2 over the base and emitter regions for contacts. A thin metallic film is then evaporated over the entire wafer and penetrates through the etched openings to contact the base and emitter areas. After this film is removed from the SiO_2—it is permitted to remain in the base and emitter areas—it is then alloyed to the silicon, and connection wires are bonded to the metallized contacts (Fig. 1-9E).

Because its chip surface is completely covered with SiO_2, this type of transistor has lower leakage than the mesa transistor. Furthermore, resistors, diodes, other transistors, and like parts may be diffused into the same chip and interconnected by a metallization pattern on top of the SiO_2 to form an integrated circuit (IC) as shown in Fig. 1-12.

Field-Effect Transistors

The development of photolithographic diffusion and metallization technologies has made possible a new type of transistor called the field-effect transistor (FET), which is superior to the junction transistor in many applications. Because of its high input impedance, its characteristics more nearly approach those of a vacuum tube. Field-effect transistors are of two types: (1) the junction field-effect transistor (JFET) and (2) the metal-oxide silicon field-effect transistor (MOSFET).

Junction Field-Effect Transistor

The construction of a JFET is shown in Fig. 1-10A. A narrow channel of N- or P-type semiconductor material is created by diffusion between two materials of opposite type (P-type, in the example shown). In an N-type JFET the carriers are the free electrons, and in a P-type JFET the carriers are the free holes. When a potential is applied across the channel, the carriers move across the channel.

For example, a positive potential is applied to the drain and a negative potential is applied to the source, causing electrons to flow through the channel from source to drain. The amount of channel current flow and the resistance of the channel can now be decreased and increased, respectively, by applying a negative voltage to both P-region gates simultaneously. A depletion region is formed in the channel along the PN junction walls of the channel. The free electrons now have less area in which to move. Hence, the channel resistance

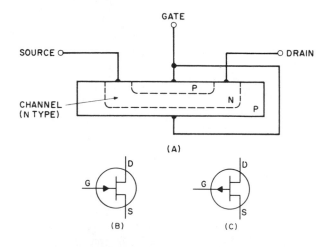

Fig. 1-10. Junction field-effect transistor (JFET), N-channel type; (A) construction, (B) N-channel schematic symbol, and (C) P-channel schematic symbol.

increases, and the source-to-drain current decreases. If the gate voltage is made sufficiently negative, the depletion regions meet, and the channel current is zero (referred to as "pinch-off"). A relatively small change in gate voltage can cause a large change in source-to-drain current, and therefore the device acts as an amplifier.

The gate-to-channel resistance of the device is very high. Being a reverse-biased diode, the JFET has characteristics similar to those of a vacuum tube. The *source* terminal is the source of carriers, corresponding to the cathode of a tube. The *drain* terminal is where the carriers are drained from the device, corresponding to the plate of a tube. The *gate* opens and closes to control the channel current, corresponding to the grid of a tube.

Metal-Oxide Silicon Field-Effect Transistor

The metal-oxide silicon field-effect transistor (MOSFET), sometimes called an insulated-gate field-effect transistor (IGFET), is of two types: (1) the depletion type, and (2) the enhancement type.

In the depletion type (Fig. 1-11) a channel of semiconductor material exists between the source and drain. The gate terminal is connected to a metal plate insulated from the channel by a very thin layer of SiO_2. Carriers are present in the channel with no bias voltage applied to the gate. A reverse bias causes carriers to be attracted toward the plate. The result is a narrowing of the channel, increased resistance of the channel, and reduced source-to-drain current flow. When a forward-bias voltage is applied, there is an opposite effect. The channel is made larger, resistance is reduced, and current is increased.

The enhancement type has only some minority carriers present in the

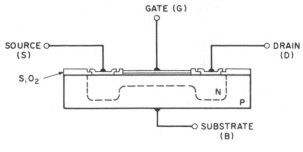

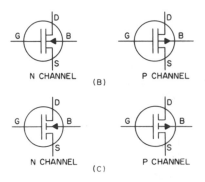

Fig. 1-11. MOSFET transistor: (A) construction; (B) depletion type; and (C) enhancement type.

channel with no bias voltage applied. A forward voltage draws carriers into the channel, decreasing channel resistance and increasing source-to-drain current. When a negative voltage is applied to the gate, carriers are repelled from the channel with the opposite result.

The MOSFET is different from the JFET and standard junction transistors in that it can operate with forward and reverse bias voltages. Also, because the gate is insulated from the channel by a SiO_2 layer, the device has an extremely high input resistance; typical values are over a thousand megohms. An additional advantage is that multiple-gate MOSFET's can be easily made. These are particularly useful in multiinput gate applications. A typical MOS logic gate is shown in Fig. 3-17.

INTEGRATED CIRCUITS

As the invention of the transistor ushered in a new era in digital electronic communications, the integrated circuit is ushering in another era. Commonly referred to as ICs, these devices use the photolithographic and diffusion techniques developed for epitaxial transistors to complete circuits on one semiconductor chip housed in one package. These techniques have reduced size, yielded closer tolerance characteristics, and decreased manufacturing costs.

ICs are now being manufactured containing hundreds of transistors at costs of less than a penny a transistor. In digital applications, ICs are used that contain entire adder, counter, memory, or calculator circuits. A typical IC may contain over 10 transistors, 20 resistors (or transistors used as fixed resistors), several capacitors, and many diodes.

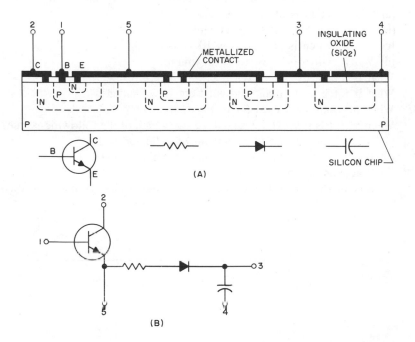

Fig. 1-12. Simple IC: (A) cross-sectional view, and (B) schematic.

The IC is constructed on one silicon chip, as shown in Fig. 1-12. All devices on the chip are isolated from each other, since each forms a reverse-biased NP junction with the P-type substrate. The devices are interconnected by the metallization. The chip is then mounted in either a TO-5 type case, or a flat-pack epoxy block. The most popular package is the dual-in-line package (DIP), which is shown in Fig. 3-19. ICs used in digital applications may have 14 or more leads connecting them to the external circuitry.

Review Questions

1. When an ideal diode is forward-biased, is current flow greater in the forward-biased direction or reverse-biased direction? Explain why.
2. When an ideal diode is reverse-biased, is current flow greater in the forward-biased direction or reverse-biased direction? Explain why.
3. What happens to the resistance of a diode when it is forward-biased? When it is reverse-biased?
4. What is the current flow called that flows under reverse-bias conditions?
5. Referring to Fig. 1-6, which of the following is true and which is false?
    (a) When switch is at A, I_B is zero.
    (b) When switch is at B, I_C is zero.
    (c) When switch is at A, E_O is 10 volts.
    (d) When switch is at B, I_E is maximum.
6. What is meant by MOSFET?
7. What is integrated circuit and what are its advantages?
8. What are transistors used for in logic circuits?
9. When an NPN transistor is said to be *on*, which of the following is true?
    (a) It is passing a minimum amount of collector current.
    (b) It is passing a maximum amount of collector current.
    (c) It acts as an open switch.
10. What is the process of interconnecting the devices that form an IC called?

Chapter 2

Binary Numbers and Coding Systems

Our number system is based on a decimal code of ten digits. This is most convenient since we have ten fingers. Computers, however, utilize the on-off switching of electronic devices and are therefore based on a binary (two-digit) numbering system for counting and computing circuits. In addition, several different numerical coding schemes are employed. For example, an octal code system with eight numbers is most often used for inputting and outputting information in a digital computer system.

THE BINARY NUMBER SYSTEM

The binary number system uses a code of numbers *0* and *1*, as compared to the decimal number system that uses a code of numbers *0, 1, 2, 3, 4, 5, 6, 7, 8,* and *9*. When we count in binary, it compares to the decimal system as follows:

Binary Number	Decimal Number
0	0
1	1
10	2
11	3
100	4
101	5
110	6
111	7
1000	8
1001	9
1010	10

It helps to read the binary numbers as zero, one, one-zero, one-one, and so forth.

Binary to Decimal Conversion

We can convert a binary number to its decimal equivalent quickly and easily using the following system. First, look at the decimal number 6,342 in terms of its units, tens, hundreds, thousands, etc.

	thousands	hundreds	tens	units
6,342 =	6000	+ 300	+ 40	+ 2

In the same way, look at the binary number 1111:

$$1111 = 1000 + 100 + 10 + 1$$

Now converting each binary number to its decimal equivalent, using the table, we get:

$$1111 = 1000 + 100 + 10 + 1$$

$$15 = 8 \quad + \quad 4 \quad + 2 \quad + 1$$

It can now be seen that each binary digit has a place value and they are:

$$\text{etc.} \quad \longleftarrow \quad 64 \quad 32 \quad 16 \quad 8 \quad 4 \quad 2 \quad 1$$

We can find the next binary digit value by taking the previous value and multiplying it by 2. In addition, the following rules must be followed in converting from binary to decimal numbers:

1. A *1* in a binary digit position means add the value of the particular decimal place value.

2. A *0* in a binary digit position means disregard its particular decimal value.

For example, find the decimal equivalent of 10101.

$$10101 = 16 + \cancel{8} + 4 + \cancel{2} + 1 = 21$$

Another example, find the decimal equivalent of 1101101.

$$1101101 = 64 + \cancel{32} + 16 + 8 + 4 + \cancel{2} + 1 = 109$$

Fractions are handled in a similar manner. The value of each digit to the right of the decimal is:

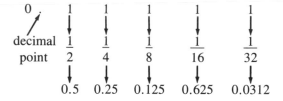

For example, find the decimal equivalent of 0.101.

$$0.101 = 0.5 + \cancel{0.25} + 0.125 = 0.625$$

Another example, find the decimal equivalent of 101.01. The numbers to the left of the decimal are handled as whole numbers and the numbers to the right of the decimal point are handled as fractions. Thus:

$$101.01 = 4 + \cancel{2} + 1 + \cancel{0.5} + 0.25 = 5.25$$

Decimal To Binary Conversion

To convert from decimal to binary numbers, we reverse the previous procedure, by breaking down a number into place-value numbers (8, 4, 2, 1, etc.). For example, 11 is expressed as follows:

$$11 = 8 + 4 + 2 + 1$$
$$1 \quad 0 \quad 1 \quad 1$$

Hence, 11 in decimal is equal to 1011 in binary.
A second example, find the binary equivalent of the decimal number 44.

$$44 = 32 + 16 + 8 + 4 + 2 + 1$$
$$1 \quad 0 \quad 1 \quad 1 \quad 0 \quad 0$$
$$44 = 101100$$

Fractions are handled in a similar manner. For example, find the binary equivalent of 0.815.

$$0.815 = 0.5 + 0.25 + 0.125 + 0.0625$$
$$0. \quad 1 \quad 1 \quad 0 \quad 1$$
$$0.815 = 0.1101$$

Binary Addition

We can add binary numbers in a manner similar to adding decimal numbers. There are four basic rules:

rule 1,	*rule 2,*	*rule 3,*	*rule 4,*
0	1	1	11
+0	+0	+1	+11
0	1	10	110

When we add binary numbers we carry to the next column as required. Example,

decimal	*binary*
5	101
+6	+110
11	1011

Yet another example,

23	10111
+7	+00111
30	11110

Binary Subtraction

Subtraction of binary numbers is based on four basic rules:

$$
rule\ 1,\ \begin{array}{r} 0 \\ -0 \\ \hline 0 \end{array} \qquad
rule\ 2,\ \begin{array}{r} 1 \\ -0 \\ \hline 1 \end{array} \qquad
rule\ 3,\ \begin{array}{r} 1 \\ -1 \\ \hline 0 \end{array} \qquad
rule\ 4,\ \begin{array}{r} 10 \\ -01 \\ \hline 01 \end{array}
$$

We subtract binary numbers in a similar manner to decimal numbers; that is, column by column and borrowing from the next column as required. Example,

$$
\begin{array}{cc}
decimal & binary \\
10 & 1010 \\
-6 & -0110 \\
\hline
4 & 0100
\end{array}
$$

Another example,

$$
\begin{array}{cc}
14 & 1110 \\
-9 & -1001 \\
\hline
5 & 0101
\end{array}
$$

Negative numbers are handled in the same manner as with decimal numbers. The sum will have the sign of the larger number. Example,

$$
\begin{array}{cc}
5 & 0101 \\
-9 & -1001 \\
\hline
-4 & -0100
\end{array}
$$

Complements of Binary Numbers

The foregoing method of subtraction is seldom used. A more efficient method, utilizing complements of binary numbers is used since it requires much less circuitry. To do this we will use what is called the 1's and 2's complements of a binary number.

Each binary number is composed of binary digits. Each binary digit is usually referred to as a *bit*. For example, we say that the binary number 1001 is composed of 4 bits and 1010101 has 7 bits.

The 1's complement of a binary number is found by changing each bit to its complement; in other words, by changing each 1 to a 0 and each 0 to a 1. Example,

the 1's complement of 101 is 010

the 1's complement of 1101 is 0010

The 2's complement is found by adding 1 to the 1's complement. Example,

the 2's complement of 101 is 010 + 1 = 011

the 2's complement of 1101 is 0010 + 1 = 0011

the 2's complement of 001 is 110 + 1 = 111

Subtraction using a complement is performed by taking the 2's complement of the number being subtracted, adding the two numbers and disregarding the last carry. Example,

	decimal	*binary*	*2's complement*
	10	1010	1010
	− 6	−0110	+1010 (2's complement of 0110)
	4	0100	10100

disregard last carry

Another example,

	14	1110	1110
	− 9	−1001	+0111
	5	0101	10101

When the carry is 1, the answer is positive. When there is no carry, the answer is negative and is the 2's complement of the result.

The End-Around Carry

Using complements reduces subtraction to an operation in addition. Hence, the computer's circuitry is designed only to add, making the system considerably simpler than otherwise. Another method, is to use the 1's complement and add the last carry to the resultant to get the answer. This is called the *end-around carry*. Example,

```
  10            1010
  −6          +1001   (1's complement of 0110)
   4          10011
              
               0011
            +     1   (end-around carry)
              0100 answer
```

When there is a carry the answer is positive, and in binary form. When there is no carry, the answer is negative and in the 1's complement of the resultant. Example,

	decimal	*binary*	*1's complement*
	5	0101	0101
	−9	−1001	−0110
	−4	−0100	01011

no carry   −0100   (1's complement)

Binary Multiplication

Binary multiplication is similar to decimal multiplication. The following rules apply:

rule 1, 0	*rule 2,* 0	*rule 3,* 1	*rule 4,* 1
×0	×1	×0	×1
0	0	0	1

Example,

$$
\begin{array}{r}
9 \\
\times 5 \\
\hline
45
\end{array}
\qquad
\begin{array}{r}
1001 \\
\times 101 \\
\hline
1001 \\
0000 \\
1001 \\
\hline
101101
\end{array}
$$

Binary Division

Division follows the same procedure as in decimal division. Example,

$$
\begin{array}{r}
3 \\
4\overline{)\ 12}
\end{array}
\qquad
\begin{array}{r}
11 \\
100\overline{)\ 1100} \\
\underline{100} \\
100 \\
\underline{100}
\end{array}
$$

OCTAL NUMBERS

In an electronic computer, decimal numbers must be converted to binary numbers. This conversion process can be made simpler and hence require less circuitry if the decimal numbers are first converted to octal numbers (base 8) and then converted into digital bits.

The octal numbers are 0, 1, 2, 3, 4, 5, 6, 7, 10, 11, 12, 13, 14, 15, 16, 17, 20, 21, 22, 23, 24, 25, 26, 27, etc. The numbers 8, 9, 18, 19, 28, 29, etc. are not used. After reaching 7 the next number is 10, after 17, the next number is 20, and so on. Each digit position is an octal number corresponds to a power of 8, as follows:

$$
\text{etc.} \longleftarrow\quad 8^3 \quad 8^2 \quad 8^1 \quad 8^0 \qquad 8^{-1} \quad 8^{-2} \quad 8^{-3} \quad\longrightarrow\quad \text{etc.}
$$

To convert from decimal to octal numbers, divide the decimal number by 8, noting the remainder. Then successively divide each remainder by 8 noting the remainder after each division. Take the remainders in reverse order to form the octal number.

Example, Find the octal number for decimal 21.

$$
\left.\begin{array}{l}
21 \div 8 = \ 2 \quad \text{remainder} = 5 \\
2 \div 8 = \ 0 \quad \text{remainder} = 2
\end{array}\right\} = 25 \ (\text{octal number})
$$

Another example, convert 252 to an octal number.

$$
\left.\begin{array}{l}
252 \div 8 = 31 \quad \text{remainder} = 4 \\
31 \div 8 = \ 3 \quad \text{remainder} = 7 \\
3 \div 8 = \ 0 \quad \text{remainder} = 3
\end{array}\right\} = 374 \ (\text{octal number})
$$

To convert a fraction, multiply the fraction by 8, noting the whole number value of the product and successively multiplying the fractional part of the product. The whole numbers then, taken in order, are the octal number.

Example, convert 0.19 to an octal number.

$$0.19 \times 8 - 1.52 \qquad 1$$
$$0.52 \times 8 - 4.16 \qquad 4$$
$$= 0.1412 \text{ (octal number)}$$
$$0.16 \times 8 - 1.28 \qquad 1$$
$$0.28 \times 8 - 2.24 \qquad 2$$

We could continue the operations if more accuracy were required. To convert octal numbers to decimal numbers, multiply the octal digit by its place value and add its products. The place values are

etc. $\longleftarrow$ 512 64 8 1 $\quad$ $1/8$ $\quad$ $1/64$ $\quad$ $1/512$ $\longrightarrow$ etc.

octal point

Each whole octal value is found by multiplying the preceeding value by 8 and adding the products. Each fractional octal value is found by multiplying the preceeding value by $1/8$.

Example, convert octal 17 to its decimal value.

$$(1 \times 8) + (7 \times 1) = 8 + 7 = 15 \text{ (decimal)}$$

Another example, convert octal 365 to its decimal equivalent.

$$(3 \times 64) + (6 \times 8) + (5 \times 1) = 192 + 48 + 5 = 245 \text{ (decimal)}$$

Octal-to-Binary Conversion

We convert from octal to binary and vice versa as follows:

Octal	Binary
0	0
1	1
2	10
3	11
4	100
5	101
6	110
7	111

To convert larger numbers we convert one digit at a time.

Example, convert octal 27 to binary.

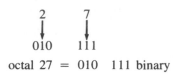

octal 27 = 010   111 binary

Another example, convert octal 2765 to binary.

2   7   6   5
↓   ↓   ↓   ↓
010   111   110   101

octal 2765 = 010   111   110   101   binary

Binary to octal conversion merely reverses the process.

Example, convert binary 001 101 011 to octal.

001   101   011
↓   ↓   ↓
1   5   3

binary 001 101 011 = 153 octal

The following example shows how octal numbers make decimal-to-binary conversion of large numbers a simpler task than direct conversion.

Example, convert decimal 434 to binary using octal numbers.

$$434 \div 8 = 54 \quad \text{remainder} = 2$$
$$54 \div 8 = 6 \quad \text{remainder} = 6 \quad \Big\} \text{ octal} = 662$$
$$6 \div 8 = 0 \quad \text{remainder} = 6$$

6   6   2
↓   ↓   ↓
110   110   010

Therefore, decimal 434 = octal 662 = binary 110 110 010

If we were to attempt to convert directly from decimal to binary, it would take nine steps. Using octal numbers reduces the conversion to six steps.

BINARY CODES

In computers, calculators, and digital instruments, a number of different codes are used which are a compromise between the decimal and binary number systems. They are referred to as *binary-coded decimal codes* (BCD). There are many widely used BCD codes which are described in the following paragraphs.

A BCD code is one in which the decimal numbers are encoded, one at a time, into groups of binary numbers. The binary groups may consist of 4 bits, 5 bits or 6 bits, etc.

8421 Code

The 8421 code is a 4-bit code. The values are 8, 4, 2, and 1, the same as in the binary number system, up to the number 9. A decimal number is changed to its equivalent 4-bit binary number as shown in the following example:

5	3	2
↓	↓	↓
0101	0011	0010

Hence, 0101 0011 0010 is the 8421 code for the decimal 532. The following table shows some of the 8421 code.

Decimal	8421	Binary
0	0000	0000
1	0001	0001
2	0010	0010
3	0011	0011
⋮	⋮	⋮
8	1000	1000
9	1001	1001
10	0001 0000	1010
11	0001 0001	1011
12	0001 0010	1100
⋮	⋮	⋮
98	1001 1000	1100010
99	1001 1001	1100011
100	0001 0000 0000	1100100
101	0001 0000 0001	1100101

Note that 1001 is the largest 4-bit group and therefore only 10 of the possible sixteen 4-bit groups are used. The numbers 1010 through 1111 are not used. Should any of these numbers occur in a system using the 8421 code, we would know that an error has occurred. Note also that up to 9, the binary and 8421 codes are the same. Above 9 they differ.

The advantage of the 8421 code is that it is easy to convert to and from decimal numbers because we can encode, or decode, one digit at a time. The disadvantage is that the rules of binary addition apply only to the individual groups, but not to the whole number. Usually, when a manufacturer states that a BCD code is used in the machine, he means the 8421 code.

Excess-3 Code

The excess-3 code is developed by adding 3 to each decimal digit before converting to binary.

Example, convert 15 to an excess-3 binary number.

Step 1	1	5
	+3	+3
	4	8
Step 2	4	8
	↓	↓
	0100	1000

Therefore, 15 equals 0100 1000 in the excess-3 binary code.

The excess-3 code also uses only 10 of the 16 possible 4-bit codes. If any of the unused groups turn up as an answer, we know that an error has occurred in the computer system. The advantage of the excess-3 code over the 8421 code is that ordinary binary addition can be used. The 1's and 2's complements can be used to accomplish subtraction.

The Gray Code

The Gray code is widely used in input-output devices, analog-to-digital converters, and other peripheral equipment. In the Gray code, each Gray number differs from the preceding number by a single bit. The Gray code however is not suitable for arithmetic operations and therefore we find that the Gray code number is converted to a binary number for the arithmetic operation and then converted back to a Gray number for output. The following table shows the Gray versus binary numbers up to 14.

Decimal	Gray Code	Binary
0	0000	0000
1	0001	0001
2	0011	0010
3	0010	0011
4	0110	0100
5	0111	0101
6	0101	0110
7	0100	0111
8	1100	1000
9	1101	1001
10	1111	1010
11	1110	1011
12	1010	1100
13	1011	1101
14	1001	1110

The Word and the Parity Bit

A group of bits which are stored or shifted, as a group, in the computer, are referred to as a *word*. During the movement of words from memory to the arithmetic unit or vice versa, errors can occur. For example, one of the 1's may be accidentally changed to a 0 by noise or a circuit fault. This could result in the output being incorrect. A method is therefore required to detect such an error. This is most often accomplished by adding a *parity bit* to the word.

Both *even* and *odd* parity bits are used. In the even parity system, if a binary number has an odd amount of 1's, a 1 parity bit is attached to the word. Conversely, if an even number of 1's exist in the word, a zero is attached to the word. The odd-parity bit is just the opposite of the even-parity bit.

This system will detect a "one-bit" error. The likelihood of an error occurring of more than one bit in a word is minimal in most equipment. However, equipment such as magnetic tape transports are more subject to bit error, and often a double-parity check is used.

Review Questions

1. Convert the following decimal numbers to binary numbers.
   (a) 26
   (b) 52
   (c) 186
2. Convert the following binary numbers to decimal numbers.
   (a) 1011
   (b) 10101
   (c) 1110101
3. Add the following numbers using binary number addition.
   (a) 9 + 7
   (b) 37 + 18
   (c) 56 + 78
4. Subtract the following numbers using binary numbers.
   (a) 9 − 7
   (b) 37 − 18
   (c) 78 − 56
5. Using the 1's and 2's complements, do the following.
   (a) 12 − 9
   (b) 42 − 31
   (c) 92 − 69
6. Convert the following decimal numbers to octal numbers.
   (a) 13
   (b) 29
   (c) 132
7. Convert the following decimal numbers to binary by first converting to octal.
   (a) 521
   (b) 932
   (c) 8,062

8. Encode the following numbers into 8421 BCD numbers.
    (a) 62
    (b) 721
    (c) 1,041
9. Decode the following 8421 BCD numbers.
    (a) 0101 1100
    (b) 0111 0100 1000
    (c) 1010 0011 0000 1001
10. Encode the following numbers into the excess-3 code.
    (a) 61
    (b) 724
    (c) 2,632
11. Attach an even-parity bit to the following words.
    (a) 0101 1100
    (b) 0111 0100 1000
    (c) 1010 0110 0000 1001
12. If the following binary numbers had the even-parity bits shown, decide whether the numbers are correct or whether they contain an error.
    (a) 1100 0101 1
    (b) 0100 1000 0111 1
    (c) 0011 0000 1010 1001 0

Chapter 3

Digital
Logic Gates

The circuits which perform the binary operations discussed in Chap. 2 are referred to as *logic circuits*. Since these circuits switch between the 0 and 1 states, they are like switches that open and close and are therefore referred to as *gates*. A gate has two or more inputs and one output. This output occurs when certain conditions are met. The operation of the gate may be analyzed through the use of a *truth table* which shows all the input and output possibilities of the logic gate circuit.

The very earliest gate circuits used relays. Later, vacuum tubes were used. Following the introduction of solid-state devices, diodes and transistors were used. Presently, gates are almost exclusively found in integrated circuit (IC) form. Therefore, it will be better to concern ourselves with the gate as a functional block rather than being concerned with the inner workings of the circuit.

THE "OR" AND "NOR" GATES

The OR gate produces a 1 (hi) when one *or* more of its inputs are 1 (hi). When all inputs are 0 (lo) the output is 0 (lo). A dual-input OR gate circuit using resistors and transistors is shown in Figure 3-1A.

With zero voltage on the base of both transistors, 0 inputs (lo), both transistors are biased-off, and zero voltage appears at the output. This is a 0 (lo) output. Actually, some leakage current will flow through the transistor and a slight positive voltage will be present. However, it will be so low that we can consider it as being zero voltage.

If a positive voltage, of sufficient level to saturate the transistor, is at the A input, transistor Q_1 is turned on and $+V_{cc}$ appears at the output terminal. Actually, there is a slight drop across the collector-to-emitter resistance of the transistor. However, it is so small that we can neglect it. The positive voltage at the A input is a 1 level (hi), and the $+V_{cc}$ voltage appearing at the output is also a 1 level.

If the B input is 1, and the A input is 0, transistor Q_2 will turn on, and the output will be 1. If both A and B inputs are 1, both transistors turn on, and the output will also be a 1. The functional logic symbol for an OR logic gate is shown in Fig. 3-1(B). The operation of the OR gate may be summarized in a *truth table*. The truth table for an OR gate is shown in Fig. 3-1(C). The truth

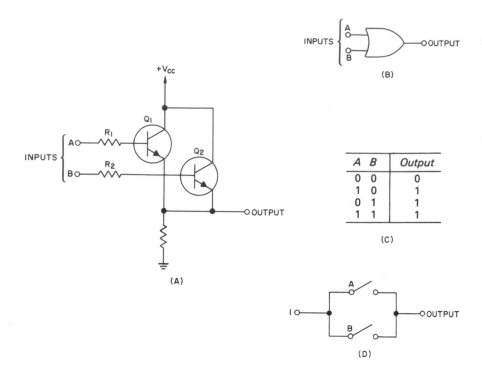

Fig. 3-1. The OR gate: (A) circuitry, (B) logic symbol, (C) truth table, and (D) OR gate equivalent using switches.

table indicates all possible input and output states of the gate. The equivalent of an OR gate using switches is shown in Fig. 3-1(D), as two switches in parallel. If either switch, or both switches, are closed the output will be 1. Only when both switches are open will the output be 0.

If the output is taken from the collector, instead of the emitter, the output will be the complement of the OR gate and is referred to as a *NOR* gate. Such a circuit is shown in Fig. 3-2(A). With all inputs at 0, both transistors are off and the output will be 1 ($+V_{cc}$). When any input is at a 1 level, the respective transistor is turned on and the output will be 0. Notice that the logic symbol, shown in Fig. 3-2(B), has a small circle at the output of the gate to indicate that the NOR output is the complement of the OR gate.

THE AND AND NAND GATES

The AND gate, shown in Fig. 3-3, has a 1 output only when all inputs are 1. It can be considered as an "all-or-nothing" gate. With no voltage (0 level) on the base of either transistor, both are biased-off and zero voltage (0 level) appears at the output. A positive voltage on one base forward-biases that transistor.

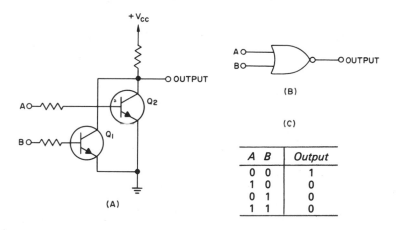

Fig. 3-2. A 2-input NOR gate: (A) circuitry, (B) logic symbol, and (C) truth table.

However, the other transistor is still biased-off and the output remains at 0 level. When a positive voltage is put on the base of both transistors, both are turned on and $+V_{cc}$ (1 level) will appear at the output.

The circuit is called an AND gate since *only* when A and B inputs are 1 will the output be 1. The circuit, as shown in Fig. 3-3(D) is effectively two switches connected in series, so that switches A *and* B must be closed to obtain a 1 output. The logic function symbol for an AND gate is shown in Fig. 3-3(B). The truth table for the AND gate is shown in Fig. 3-3(C).

Like the NOR gate, if the output is taken from the collector instead of the emitter circuit, the output will be the complement of the AND gate and is referred to as a *NAND* gate. When *all* the inputs are 1, the output will be 0. When any, or all the inputs are 0, the output will be 1.

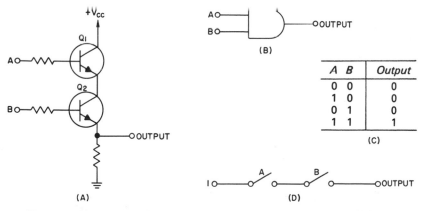

Fig. 3-3. A 2-input AND gate: (A) circuitry, (B) logic symbol, (C) truth table, and (D) equivalent circuit using switches.

THE INVERTER

The inverter, sometimes called a *NOT* circuit, as shown in Fig. 3-4, has one input and one output. It merely inverts the input signal. If the input is a 1, the output will be a 0 and vice versa. The output is the complement of the input.

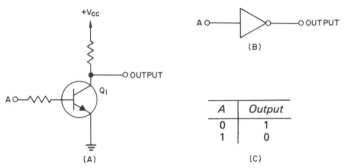

Fig. 3-4. The inverter or NOT circuit: (A) circuitry, (B) logic symbol, and (C) truth table.

POSITIVE AND NEGATIVE LOGIC SYSTEMS

It is important to know whether a logic system is based on the 1 level being positive or negative with respect to the 0 level. It is important because an AND gate in a positive logic system operates as an OR gate in a negative logic system. Also, an OR gate in a positive logic system operates as an AND gate in a negative logic system.

To understand this, consider a gate where the voltage levels are 0 volts and 5 volts, as shown in Table 3-1(A). In a positive logic system, where 0 = 0 volts and 1 = 5 volts, the gate will be an OR gate as shown in Table 3-1(B). However, in a negative logic system, where 0 = 5 volts and 1 = 0 volts, the same gate will be an AND gate, as shown in Table 3-1(C).

Table 3-1. An OR gate becomes an AND gate when we change from positive to negative logic systems.

(A) Voltage Levels			(B) Positive Logic OR Gate			(C) Negative Logic AND Gate		
A	*B*	*Output*	*A*	*B*	*Output*	*A*	*B*	*Output*
0V	0V	0V	0	0	0	1	1	1
0V	5V	5V	0	1	1	1	0	0
5V	0V	5V	1	0	1	0	1	0
5V	5V	5V	1	1	1	0	0	0

THE UNIVERSAL GATE

It is possible to accomplish all the previous gating functions using only NAND or NOR gates. This permits great economies in manufacturing, since a digital system can be built using only one type gate to perform all logic gating functions.

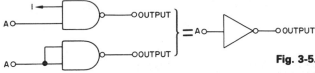

Fig. 3-5. Using a NAND gate to make an inverter.

For example, we can make an inverter with a NAND gate by connecting all inputs together, or by connecting all but one input to a 1 level, as shown in Fig. 3-5. It is also possible to make an AND gate from a NAND gate by inverting the NAND gate's output. This is shown in Fig. 3-6.

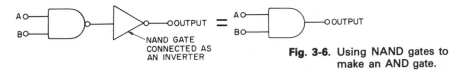

Fig. 3-6. Using NAND gates to make an AND gate.

Figure 3-7 illustrates how either NAND or NOR gates are often used to create any of the six basic logic circuit functions. (Note: the exclusive OR gate will be discussed in Chapter 6.) For example, an OR gate can be made using

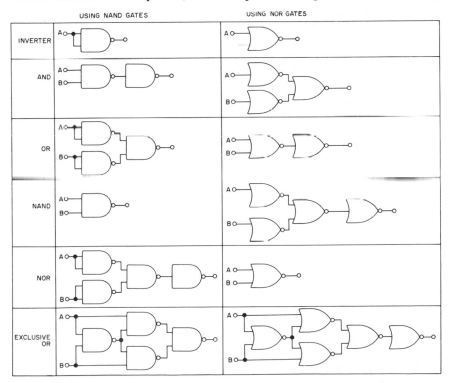

Fig. 3-7. Using NAND and NOR gates to perform basic logic functions.

three NAND gates or two NOR gates. When using the NAND gates, the first two NAND gates function as inverters so that with A = 0 and B = 0, the output will be 0. If either A or B = 1, the output will be 1.

DECIMAL-TO-BINARY ENCODING

Since a computer works in binary, it is necessary to convert our decimal numbers to binary numbers to enter them into the computer. A logic circuit for this conversion is shown in Fig. 3-8. The decimal inputs might be keyboard switches for the decimal numbers. The switches are normally connected to 0. When a key is depressed, the related switch goes from a 0 to a 1 level. For example, when the 7-key is depressed, a 1 level is put on one of the inputs of the 1, 2, and 4-OR gates, causing a 1 output at these OR gates; the 8-OR gate will be 0 since both of the inputs are 0. The binary output will therefore be 0111, the BCD code for decimal 7.

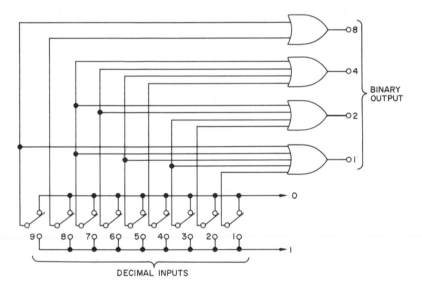

Fig. 3-8. A decimal-to-binary encoder.

BINARY-TO-DECIMAL DECODING

After the computer has completed its operation, it is usual to decode from binary to decimal so that the output appears in decimal form. A logic circuit for binary-to-decimal decoding is shown in Fig. 3-9. The circuit employes four inverters and ten AND gates. The outputs of the inverters are labeled $\overline{1}, \overline{2}, \overline{4}$, and $\overline{8}$ to indicate that they are complements of 1, 2, 4, and 8.

The operation of the decoder is such that a particular binary number will cause its respective decimal gate to produce a 1 output. For example, the binary number 0110 (decimal 6) will cause the 1, 8, $\overline{2}$, and $\overline{4}$ lines to have a 1 level, while the 2, 4, $\overline{1}$, and $\overline{8}$ lines will be at a 0 level. The decimal-6 gate will therefore have all 1 inputs, and all the other gates will have at least one 0 input. The result will be that only the decimal-6 gate will have a 1 output, and the binary input will have been converted to a decimal output.

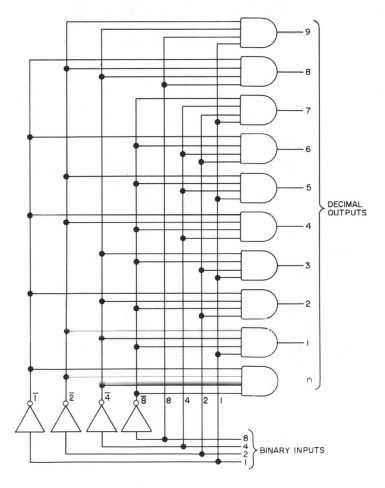

Fig. 3-9. A binary-to-decimal decoder.

TYPES OF GATE CIRCUITS

Gating circuits may be made using combinations of diodes, transistors, and resistors in a wide variety of circuit types. Each circuit type has advantages and disadvantages compared with the other types. The following is a very brief discussion of types of circuits in use today and their relative features.

DL Logic

Diode logic (DL) employs diodes and resistors. A 2-input DL OR gate is shown in Fig. 3-10(A). When either the A or B input is high, its respective diode is forward-biased, and the input voltage appears at the output (less the forward voltage drop across the diode). This type of gating has the disadvantage that the forward voltage drop of the diodes, over several gating stages, is cummulative and diminishes the difference between the logical 1 and 0 levels. It is therefore necessary to insert a transistor switching circuit, as required, to restore the 1 and 0 levels.

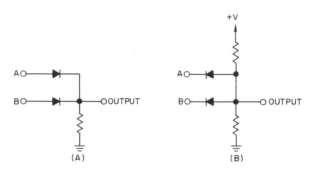

Fig.3-10. Diode logic (DL) gating circuits: (A) OR gate, and (B) AND gate.

In the DL AND gate, shown in Fig. 3-10(B), the 0 input to each diode forward-biases the diodes putting the output at a 0 level. Only when both inputs are at the 1 level are both diodes turned off and +V appears at the output.

Diode logic is very popular in simple logic systems such as the decimal-to-binary encoder. Such an encoder is shown in Fig. 3-11. The operation can be seen if the 5-switch is closed. The two diodes connected to this line will be forward-biased (low resistance), lowering the potential on outputs D and B to the 0 level (0.6 volt) while outputs C and A remain at the 1 level (approximately 5 volts). The output will therefore be 0101, which is the binary equivalent of decimal 5.

RTL Logic

Resistor-transistor logic (RTL) has the advantage of maintaining the difference between the 0 and 1 levels. This is because the transistor is always switching between cutoff and saturation, and hence "recreates" the pulse. The examples shown in Figs. 3-1, 3-4, and 3-5, at the beginning of this chapter, are RTL-type gating circuits.

DTL Logic

Diode-transistor logic (DTL) is actually DL logic with a transistor switch following each gate to overcome the disadvantage of the DL logic circuit. A

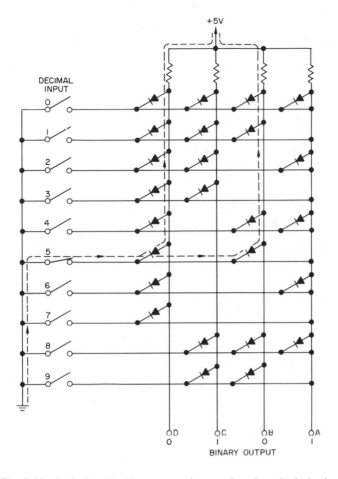

Fig. 3-11. A decimal-to-binary encoder matrix using diode logic.

typical DTL logic ciruit is shown in Fig. 3-12. Compared to RTL, DTL has the advantage of greater noise immunity. However, it costs more than RTL.

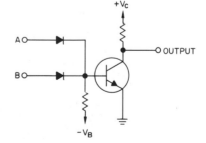

Fig. 3-12. A DTL NOR gate.

DCTL Logic

Direct-coupled transistor logic is essentially an RTL circuit where transistors are used instead of base resistors. This type of logic circuit is often used in integrated circuits (ICs), since in an IC, a transistor usually occupies less space on the chip. A typical DCTL logic circuit is shown in Fig. 3-13.

TTL Logic

Transistor-transistor logic (TTL) is very popular for use in ICs because multiinput gates are easily and more economically produced. Further, TTL has a high noise immunity (typically 1 volt). A typical TTL type NAND gate is shown in Fig. 3-14. When any or all of the inputs are at 0 level, transistor Q_1 is forward-biased, putting 0 volts on the base of transistor Q_2. Q_2 is biased-off, and the output will be a logical 1. When all the inputs are at a 1 level, the base-emitter junction of Q_1 is biased-off, but the base-collector junction is forward-biased, putting a positive voltage on the base of transistor Q_2. Q_2 is then forward-biased, yielding a 0 output.

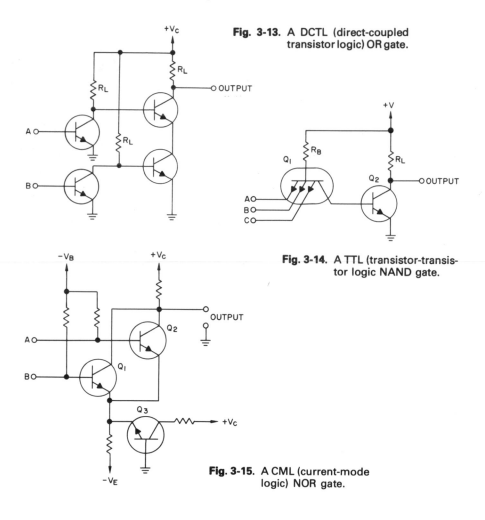

Fig. 3-13. A DCTL (direct-coupled transistor logic) OR gate.

Fig. 3-14. A TTL (transistor-transistor logic NAND gate.

Fig. 3-15. A CML (current-mode logic) NOR gate.

CML Logic

Current-mode logic (CML) is used where very fast switching speeds and high noise immunity are required. A CML NOR gate is shown in Fig. 3-15. Transistor Q_3 provides a constant-current source which prevents transistors Q_1 and Q_2 from conducting to saturation. This eliminates the delay inherent in saturating the gates, due to the need to sweep all the carriers out of the base region before the transistor can begin to respond to the fall-off of the input pulse.

MOS Logic

The metal-oxide silicon (MOS) field-effect transistor (FET) is used, particularly in medium- and large-scale ICs. This is because MOS devices take up much less space on the IC chip and consume less power, which in turn means that more devices can be put on the chip and more circuits can be included in one integrated circuit. Using MOS technology, all the logic circuity for calculator has been put on one IC chip (refer to Chap. 12). The disadvantage of MOS devices is their speed limitations, due to the inherent gate capacitance of the device.

A typical MOS NAND gate is shown in Fig. 3-16. The switching transistor Q_1 has three gates. All three gates must be turned on for the device to pass current. Transistor Q_2 acts as a load resistor for the circuit since an MOS transistor occupies considerably less space than a resistor on the chip. A fixed voltage on the gate of Q_2 establishes the source-to-drain resistance of Q_2.

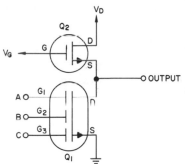

Fig. 3-16 A MOSFET 3-input NAND gate.

FAN-IN AND FAN-OUT

The *fan-in* of a logic gate is the number of inputs that a gate has. For example, the gate shown in Fig. 3-16 has a fan-in of three. The number of circuits that can be driven by the output of the gate determines its *fan-out* rating. Since junction transistors require input current to operate, they represent a load to the driving circuit. If a gate has a fan-out rating of five, then the devices connected to its output should not have a total fan-in of more than five. If this rating is exceeded, there is a likclihood that not all the driven circuits will operate.

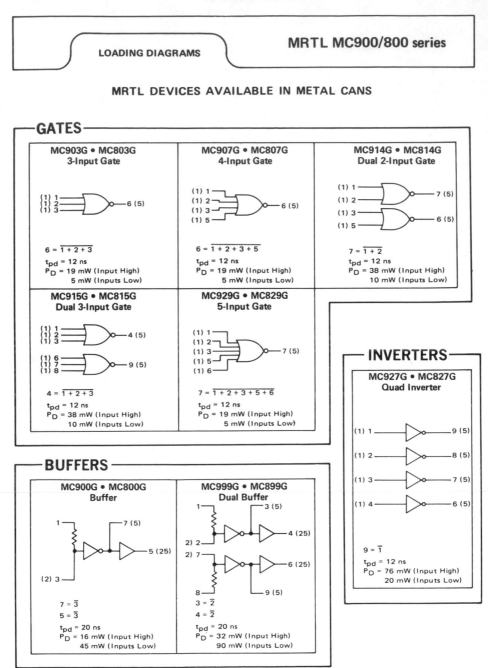

Fig. 3-17. A typical logic family—RTL medium power. (*Courtesy Motorola*)

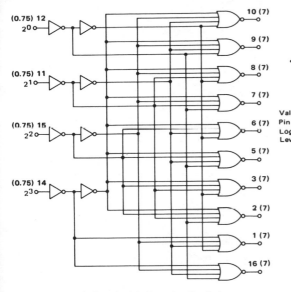

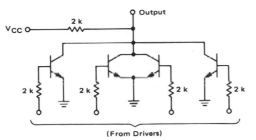

Number in Parenthesis Indicates Loading Factor.

BCD-TO-DECIMAL DECODER

MC770P · MC870P*

*P suffix = 16 pin dual-in-line plastic package, Case 612.

TRUTH TABLE

INPUT (BCD)				OUTPUT (DECIMAL)									
2^3	2^2	2^1	2^0	0	1	2	3	4	5	6	7	8	9
14	15	11	12	10	9	8	7	6	5	3	2	1	16
0	0	0	0	1	0	0	0	0	0	0	0	0	0
0	0	0	1	0	1	0	0	0	0	0	0	0	0
0	0	1	0	0	0	1	0	0	0	0	0	0	0
0	0	1	1	0	0	0	1	0	0	0	0	0	0
0	1	0	0	0	0	0	0	1	0	0	0	0	0
0	1	0	1	0	0	0	0	0	1	0	0	0	0
0	1	1	0	0	0	0	0	0	0	1	0	0	0
0	1	1	1	0	0	0	0	0	0	0	1	0	0
1	0	0	0	0	0	0	0	0	0	0	0	1	0
1	0	0	1	0	0	0	0	0	0	0	0	0	1
1	0	1	0	0	0	0	0	0	0	0	0	0	0
1	0	1	1	0	0	0	0	0	0	0	0	0	0
1	1	0	0	0	0	0	0	0	0	0	0	0	0
1	1	0	1	0	0	0	0	0	0	0	0	0	0
1	1	1	0	0	0	0	0	0	0	0	0	0	0
1	1	1	1	0	0	0	0	0	0	0	0	0	0

Value
Pin No.
Logic Level

t_{pd} = 36 ns
P_D = 100 mW typ (All inputs high)

4-INPUT "NOR" GATE (1-OF-10)

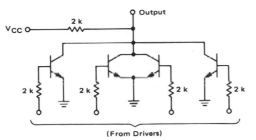

DUAL SERIES INVERTING DRIVER (1-OF-4)

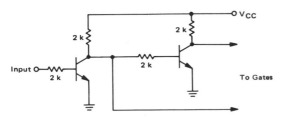
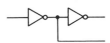

Fig. 3-18. A complete BCD-to-decimal decoder in one IC package. (*Courtesy Motorola*)

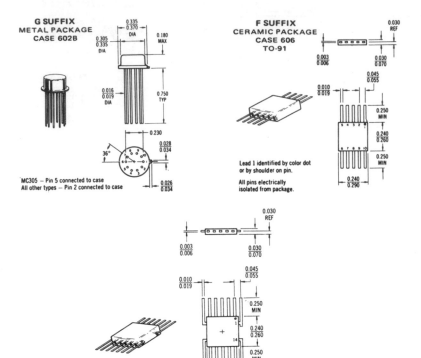

G SUFFIX
METAL PACKAGE
CASE 602B

MC305 — Pin 5 connected to case
All other types — Pin 2 connected to case

F SUFFIX
CERAMIC PACKAGE
CASE 606
TO-91

Lead 1 identified by color dot
or by shoulder on pin.

All pins electrically
isolated from package.

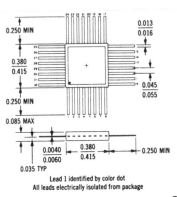

Lead 1 identified by color dot or by elbow on
lead. All leads electrically isolated from package.

F SUFFIX
CERAMIC PACKAGE
CASE 607
TO-86

Lead 1 identified by color dot
All leads electrically isolated from package

F SUFFIX
CERAMIC PACKAGE
CASE 618

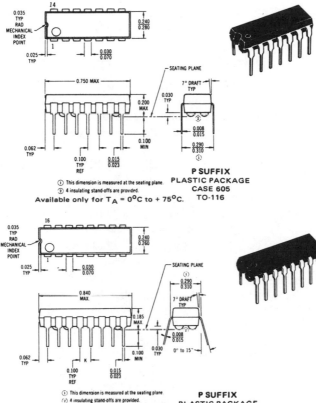

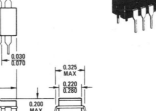

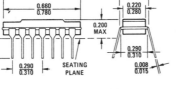

Fig. 3-19. Typical integrated circuit packages used for digital logic gates. (*Courtesy Motorola*)

DTL type circuits have high fan-in and fan-out ratings. However, this increased rating is at the expense of slower switching speed.

INTEGRATED CIRCUIT GATES

Almost all present equipment using digital logic circuitry employ IC gates because of the lower device cost and reduced manufacturing costs of IC circuits compared to discrete-component circuitry. IC manufacturers offer IC logic gates in a series usually referred to as a *family*. For example, an RTL family or a TTL family.

The RTL family of logic is generally available in high and low power ratings. The difference between these families is the values of their resistors. The high-power ICs have smaller resistors to reduce the RC time constant of the circuit and reduce the time delay inherent in the circuit. However, the reduced resistance means that the gates consume more current and have reduced fan-in and fan-out ratings. The high- and low-power RTL devices may be interfaced if the fan-in and fan-out loading rules are observed.

The DTL and TTL logic families are similar and may be intermixed (interfaced). These two logic families have high noise immunity ratings. The noise immunity rating of TTL- and DTL-type logic is typically 1 volt, while that of RTL is typically 300 mv.

A typical family of gates is shown in Fig. 3-17. This diagram briefly summarizes the specifications of each IC. For example, the MC914G contains two (dual) 2-input NOR gates. The unit is housed in a 10-pin metal "can" type package (G suffix shown in Fig. 3-19). This series is also available in a "flat" package. The pin number is shown adjacent to each input and output. The number in parenthesis indicates the fan-in and fan-out ratings. The equation below the diagram indicates the operation of the gate; in this case (MC914G), terminal 7 will have the inversion of the 1 ($\overline{1}$) or 2 ($\overline{2}$) inputs. The t_{pd} is the propagation delay of a pulse passing through the gate. P_D is the typical power dissipated by the device.

The "buffers" are low-impedance driver circuits used to drive heavy loading circuits and minimize delay time due to capacitance loading. The buffers have more fan-out capabilities than the gates alone.

Note that the $+V_{cc}$ is connected to pin 8, and ground to pin 4 in all the ICs using the 8-pin cans. In all the ICs using 10-pin cans, V_{cc} is connected to pin 10 and ground to pin 5.

Figure 3-18 is part of the specifications for a complete BCD-to-decimal decoder in one 16-pin plastic dual-inline package (DIP). The DIP is denoted by the suffix P and is shown in Fig. 3-19.

The logic diagrams in Fig. 3-17 describe the MC900/MC800 MRTL integrated circuits available in metal cans, and permit quick selection of those circuits required for the implementation of a system design. Pertinent information such as logic equations, truth tables, typical propagation delay time (t_{pd}), typical package power dissipation (P_D), pin numbers, input loading, and fan-out is shown for each device. The package pin number is shown adjacent to the terminal end. The number in parenthesis indicates the input loading factor (when on the circuit input terminal) or load driving ability—fan-out—(when on the circuit output terminal).

The number of load circuits that may be driven from an output is determined by the output loading factor and the sum of all input loading factors for the circuits connected to that output. The summation of the input loading factors should not exceed the stated drive capability of the output. Loading data are valid over the temperature range of -55 to $-+125°C$ for the MC900 Series, and 0 to $+100°C$ for the MC800 Series, with $V_{cc} = 3.0$ v $\pm$ 10%. For the TO-99 metal can, V_{cc} is applied to pin 8, with ground connected to pin 4. For the TO-100 metal can, V_{cc} is applied to pin 10, with ground connected to pin 5.

The MC770P/870P shown in Fig. 3-18 is a monolithic BCD-to-decimal decoder consisting of eight inverters and ten 4-input NOR gates which are utilized to convert binary coded decimal (8-4-2-1) input to an output, via the appropriate one of ten output lines.

Review Questions

1. Under what conditions does an OR gate have a 1 output?
2. Write a truth table for a 3-input OR gate.
3. Under what conditions does an AND gate have a 1 output?
4. Write a truth table for a 3-input AND gate.
5. Write a truth table for a 3-input OR gate using negative logic. Compare its truth table to the 3-input AND gate in question 4.
6. Under what conditions does a NOR gate have a 1 output?
7. Under what conditions does a NAND gate have a 1 output?
8. Show how a 3-input NAND gate can be used as an inverter.
9. Show how a 3-input NAND gate can be used as a 2-input NAND gate.
10. Explain how a 3-input OR gate can be made using a NAND gate and inverters.
11. Explain how a 3-input AND gate can be made using a NOR gate and inverters.
12. Explain the operation of the decimal-to-binary encoder by explaining the operation of each gate.
13. Explain the operation of the binary-to-decimal decoder by explaining the operation of each gate.

Chapter 4

Flip-Flops

The gating circuits studied in the previous chapter all require continuous input levels to operate. They could not remember their state after the inputs were removed. In other words, gates have no memory. A memory can be provided through the use of a *flip-flop* circuit.

All flip-flops have two outputs, a Q and a $\overline{Q}$ output. Each output is the complement of the other. When the Q output is high, the $\overline{Q}$ output is low and vice versa. The outputs will remain at their respective levels until instructed to change.

In addition to storing levels, the flip-flop, also known as a bistable multivibrator, can count pulses, control circuits, synchronize operations, produce required waveshapes, and perform many other functions required in digital systems.

THE R-S FLIP-FLOP

A basic flip-flop is simply a pair of amplifiers (single-input gates, or inverters) whose outputs and inputs are cross-coupled to each other, as shown in Fig. 4-1. When power is applied, one transistor will conduct more than the other, increasing its conduction to saturation where its collector voltage will be essentially at ground potential.

In the circuit shown, assume that transistor Q_1 is the higher conducting transistor and hence its collector voltage will go to essentially zero. Q_2's base voltage will therefore be zero, and Q_2 will be cut off, raising its collector voltage to V_{cc}, causing Q_1 to be forward-biased. With Q_2 turned off and Q_1 on,

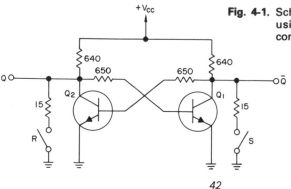

Fig. 4-1. Schematic of a simple flip-flop using amplifiers with cross-connected outputs and inputs.

the $\overline{Q}$ output will be 0 and the Q output will be 1. Now, if switch R is momentarily closed, the collector of Q_2 and the base of Q_1 will be put at zero potential, turning Q_1 off. Q_1's collector voltage will rise to V_{cc}, forward-biasing Q_2. The Q output will now be 0 and the $\overline{Q}$ output will be 1.

The R switch is used to reset the flip-flop to $Q = 0$ and $\overline{Q} = 1$ and hence is called the *reset*. The circuit will now remain reset until the S switch is momentarily closed. The S switch is called the *set* switch since it causes transistor Q_1 to turn on, and Q_2 to turn off, setting the outputs to $Q = 1$ and $\overline{Q} = 0$. The circuit will remain in the set mode until reset by the R switch.

Transistors can be used instead of the switches and the 15-ohm resistors to switch the flip-flops from logic level inputs. This arrangement is shown in Fig. 4-2. This circuit is called an *R-S flip-flop* (reset-set) or *latch*. The circuit actually consists of two 2-input NOR gates with cross-connected outputs and inputs, as shown in Fig. 4-2(B). When a 1 is at the S input, it is the same as closing the S switch in Fig. 4-1. When a 1 is at the R input, it is the same as closing the R switch.

The truth table for the R-S flip-flop and logic symbol is shown in Fig. 4-3. Note that the Q and $\overline{Q}$ outputs will remain at their respective levels until pulsed to change. Also if the Q output is already at a 1 level, a pulse at the S input will not cause a change. Conversely, if $Q = 0$, a pulse at the R input will not cause a change. If both the R and S inputs are put high, both Q and $\overline{Q}$ will be 0. Also, if R and S inputs are both put low, both the Q and $\overline{Q}$ outputs will go high for the duration of the pulse. When the pulse is removed, the flip-flop may latch to either a 1 or 0 condition, and hence will have an indeterminate output. For this reason, the inputs to an R-S flip-flop are never simultaneously pulsed.

Some manufacturers designate the Q output of the flip-flop as the "1" output and the $\overline{Q}$ output as the "0" output.

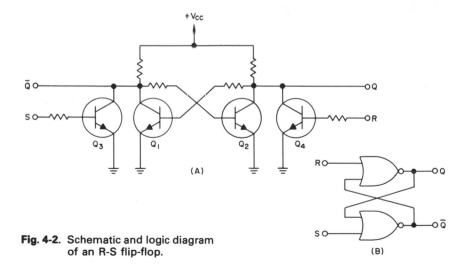

Fig. 4-2. Schematic and logic diagram of an R-S flip-flop.

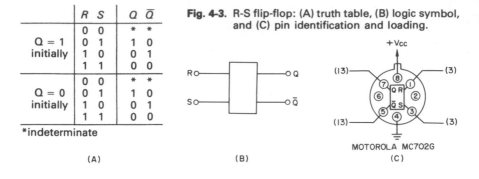

	R	S	Q	$\overline{Q}$
	0	0	*	*
Q = 1	0	1	1	0
initially	1	0	0	1
	1	1	0	0
	0	0	*	*
Q = 0	0	1	1	0
initially	1	0	0	1
	1	1	0	0

*indeterminate

(A)

(B)

(C)

Fig. 4-3. R-S flip-flop: (A) truth table, (B) logic symbol, and (C) pin identification and loading.

MOTOROLA MC702G

THE TOGGLED R-S FLIP-FLOP

The toggled R-S flip-flop, sometimes referred to as an *RST flip-flop,* has a third input, as shown in Fig. 4-4. When a pulse is applied to this input (T), the flip-flop will toggle (shift) from state to state.

If the RST flip-flop is in the state where Q = 0 and $\overline{Q}$ = 1, then the operation of the circuit is as follows. A positive pulse at the T input is inverted by transistor Q_5 and fed to diodes D_1 and D_2 through coupling capacitors. Diode D_1 is reverse-biased since the base of Q_1 has a 0 level on it. Diode D_2 is forward-biased since the base of transistor Q_2 is at a 1 level. The pulse is thus coupled through diode D_1 to the base of transistor Q_2 causing the flip-flop to toggle. On the next positive pulse at the T input, diode D_1 will be forward-biased, causing the pulse to be directed to the base of Q_1. These diodes are referred to as *steering diodes.*

Since the Q and $\overline{Q}$ outputs will go to a 1 level after two input pulses, the RST flip-flop can be said to divide-by-two. The R and S inputs are used for the same purpose as in the R-S flip-flop.

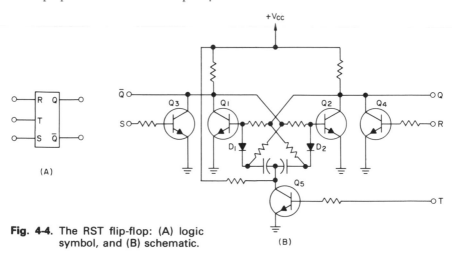

Fig. 4-4. The RST flip-flop: (A) logic symbol, and (B) schematic.

(A)

(B)

THE CLOCKED R-S FLIP-FLOP

It is often necessary to synchronize the operation of an R-S flip-flop. This is done by having a synchronizing pulse signal generated by a free-running oscillator trigger the operation of the flip-flop. This oscillator is referred to as a *clock,* and it's signal is referred to as the *clock pulse.* A synchronized R-S flip-flop, is called a *clocked R-S Flip-Flop,* and is shown in Fig. 4-5. It has a third input for the synchronizing clock pulse.

If the S input has a 1, gate A will be enabled when a 1 is present at the clock (C) input, and the flip-flop will go to the set state, if it is not already there. If the R input has a 1 level, the B gate will be enabled on the next clock pulse causing the flip-flop to reset, if it is not already

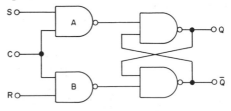

Fig. 4-5. The clocked R-S flip-flop.

there. Note that indeterminate states can occur if both the R and S inputs are 1 at the time that the clock pulse occurs at the C input.

THE TYPE-D FLIP-FLOP

The type-D flip-flop is shown in Fig. 4-6. When a toggle input is added to a clocked R-S flip-flop, it is called a *type-D flip-flop* and the toggle input is referred to as the *data input,* or D input.

Whatever level is present at the D input, prior to the clock pulse, is transmitted to the Q output of the flip-flop on the leading edge of the clock pulse. If the D input changes after the leading edge of the clock pulse has occurred, the flip-flop is not affected. A 1 at either the R or S inputs will enable their respective functions.

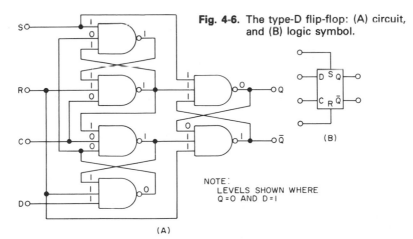

Fig. 4-6. The type-D flip-flop: (A) circuit, and (B) logic symbol.

NOTE:
LEVELS SHOWN WHERE
Q = 0 AND D = 1

(A)

THE J-K FLIP-FLOP

The J-K flip-flop is one of the most versatile flip-flops and hence is very widely used. It permits the design of counters which can count to any number desired, as will be shown in the following chapter. A J-K flip-flop logic circuit is shown in Fig. 4-7. This is a master-slave J-K flip-flop and is the type most commonly used. It is not responsive to a positive-going or steady-state clock input. It responds only to a negative going clock input that is fast changing. For this reason, the clock signal must have a very fast fall-time such as is developed by a flip-flop, a Schmitt trigger or "one-shot" multivibrator circuit (these last two circuits are discussed in Chap. 7).

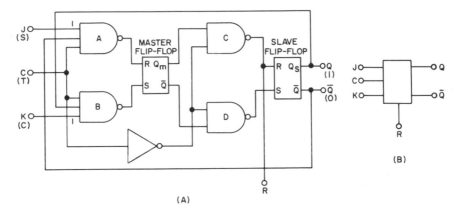

Fig. 4-7. The J-K flip-flop: (A) circuitry, and (B) logic symbol.

Referring to Fig. 4-7, the J and K input levels are passed to the master flip-flop on the leading edge of the clock pulse and held there until the trailing edge of the clock pulse allows them to be passed to the slave flip-flop. The following four possible events can occur on a negative-going clock pulse.

1. if $J = 1$ and $K = 0$, flip-flop will set.
2. if $J = 0$ and $K = 1$, flip-flop will reset.
3. if $J = 1$ and $K = 1$, flip-flop will toggle.
4. if $J = 0$ and $K = 0$, flip-flop remains in state it is in.

Typical operation of the circuit is as follows: If $J = 1$, $K = 0$ and $Q_m = 0$, the leading edge (positive-going) of the clock pulse will enable gate A with three 1 inputs. This will reset the master flip-flop causing Q_m to go to 0. The slave flip-flop will not change because the inverter provides a 0 level to gates C and D, blocking inputs to the slave flip-flop. When the trailing edge of the clock pulse occurs, gate D is enabled by 1's at its inputs and the slave flip-flop is fed a reset level, causing the Q_s output of the J-K flip-flop to go to 0.

Some manufacturer's will label flip-flop inputs and outputs differently. For example, as shown in parentheses in Fig. 4-7, the clock (C) input may be called the toggle input (T). The J input (J) may be called the set input (S). The K input may be called the clear input (C) and the reset input (R) may be called the pre-clear input (P or C_D). Also, often the Q and $\overline{Q}$ outputs are labeled 1 and 0 outputs, respectively. Table 4-1 summarizes the various flip-flop operations.

Table 4-1. Summary of Flip-Flop Operations.

Circuit Operation	Logic Diagram
R-S Flip-Flop A 1 to S input sets Q = 1 and $\overline{Q}$ = 0 A 1 to R input resets Q = 0 and $\overline{Q}$ = 1	
RST Flip-Flop Same as R-S flip-flop with added feature of a toggle (shift) input. A 1 to T input will cause flip-flop to switch states.	
Clocked R-S Flip-Flop Same as R-S flip-flop except will not set or reset until clock pulse arrives.	
Type D Flip-Flop Combines R, S, C and T inputs. The T input now called the D (data) input	
J-K Flip-Flop (Master-Slave) Responds only to fast negative-going pulses at the clock (C) input. J and K inputs must be at their levels before clock pulse arrives to operate. Assures no indeterminate states. May have direct-reset (clear) and multiple J and K inputs. Operates as shown in truth table.	

J	K	Result
1	0	set
0	1	reset
1	1	toggle
0	0	remains in state it is in

Review Questions

1. The pulse-wave train shown below is fed to the toggle input of an RST flip-flop. Sketch the Q and $\overline{Q}$ outputs on the same time-base.

2. If two clocked RST flip-flops were connected as shown below, and a 1,000-Hz pulse were fed into the T input of FF-A, what would be the frequency of the output of FF-A?
3. What would be the frequency of the output of FF-B?

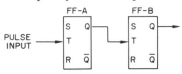

4. Show one way to make an RST flip-flop using an R-S flip-flop and two AND gates.
5. Draw the output for the J-K flip-flop circuit for the input shown (remember J-K flip-flop responds to trailing edge of pulse).

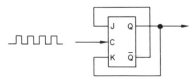

6. In a J-K flip-flop, if K = 1 and J = 0, and a clock pulse enters the C input, what levels will appear at the Q and $\overline{Q}$ outputs?
7. If K = 0 and J = 1, in a J-K flip-flop, and the clock pulse enters the C input, what levels will appear at the Q and $\overline{Q}$ outputs?
8. If, in a J-K flip-flop, the J and K inputs are at a 1 level and the clock pulse occurs at the C input, what output condition results?
9. In an RST flip-flop, if R = 1 and S = 0 when the clock pulse occurs, what condition results?
10. In an RST flip-flop, if the R and S inputs are at a 0 level when the clock pulse occurs, what condition results?
11. In a J-K flip-flop, if the J and K inputs are 0 when the clock pulse occurs, what condition results?

Chapter 5

Counters and Registers

The flip-flops discussed in Chap. 4 find their widest application in counter and register circuits. The counter circuit counts the number of pulses entering the circuit and therefore is a basic part of all computer systems. Registers are used to store binary information, such as the output of a decimal-to-BCD encoder. Registers, therefore, are a link between the main digital system and the input-output channels.

THE BASIC RIPPLE COUNTER

A 4-bit ripple, or asynchronous, counter is shown in Fig. 5-1. RST flip-flops are connected so that the Q output of the preceeding flip-flop is connected to the T input of the following flip-flop. The flip-flops toggle on the negative-going edge of the clock pulse. Flip-flop FF-A drives flip-flop FF-B, which in turn drives flip-flop FF-C, etc.

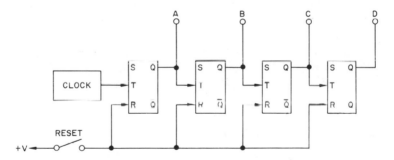

Fig. 5-1. A 4-bit ripple binary counter.

Initially, all the flip-flops are reset to Q = 0 by a positive pulse to the R terminals. FF-A will toggle to Q = 1 on the trailing edge of the first pulse, as shown in Fig. 5-2. On the trailing edge of the second pulse, FF-A will toggle to Q = 0. FF-A toggles on every clock pulse, and will therefore have an output of 1 pulse for each 2 clock pulses entering. FF-B toggles every time FF-A's output goes from 1 to 0, and therefore it will have 1 pulse out for each 2 FF-A output pulses and for each 4 clock pulses.

FF-C likewise toggles every time FF-B goes from 1 to 0 and it divides FF-B's output by 2. Therefore, FF-C will have 1 pulse for each 2 of FF-B, each

49

4 of FF-A, and each 8 clock pulses. Since FF-D is driven from FF-C's output, it will have 1 pulse out for each 2 of FF-C, each 4 of FF-B, each 8 of FF-A, and each 16 clock pulses. If outputs were taken from each Q output, they would be as shown in the truth table of Fig. 5-2.

The ripple counter, shown in Fig. 5-1, is thus a binary counter. It will count the number of clock pulses from 0 to 15 (1111) since on the 16th count the counter automatically resets to 0000. Each flip-flop stores 1 bit to make this a 4-bit binary counter having 16 distinct states. It is called an asynchronous, or ripple, counter since the pulses move through the flip-flops one at a time.

A DECADE COUNTER

The 4-bit ripple counter in Fig. 5-1 counts 16 states. However, in most readout applications we want to count 10 states. This is generally accomplished by having the counter skip steps. A simple means of doing this is to feed back pulses, as shown in Fig. 5-3. The counter will operate the same as the previous counter up through the count of 7. On the eight clock pulse FF-D will go to Q = 1, which is fed back to the set inputs of FF-B and FF-C, causing these flip-flops to set to Q = 1. The outputs of the counter will now be 1110. In other words, the counter skipped from 1000 to 1110 on the eighth clock pulse. The ninth clock pulse causes the counter output to go to 1111, and the tenth clock pulse resets the counter to 0000.

If we wish now to decode the counter output to activate a numerical display device, such as Nixie® tube (see Chap. 9), we may use the circuit shown in Fig. 5-4. The circuit consists of a 4-bit decade counter, a binary-to-decimal decoder (similar to the circuit of Fig. 3-10) driver transistors, and numerical display tube.

The counter circuit in Fig. 5-4 is slightly different from the one shown in Fig. 5-3. The counter counts up through 9 in the standard binary fashion. On the tenth pulse, however, the Q output of FF-B resets FF-D. FF-D develops a negative-going pulse at its Q output which is fed back to FF-B, resetting it to 0. Hence the output of the counter will be 0000 on the tenth clock pulse.

The decade counter circuit of Fig. 5-4 may be used to count to larger numbers by cascading decade-counter units (called DCU's for short). For example, a counter with a 999 capacity is shown in Fig. 5-5. On the tenth pulse of the counter, FF-D's Q output goes from 1 to 0, yielding a negative-going *carry* pulse to activate the next counter.

THE MODULUS OF A COUNTER

Through the proper choice of feedback, it is possible to design counters to provide any count desired. The number of possible states that a counter has is referred to as the *modulus* of the counter. For example, the counter in Fig. 5-1 has 16 possible states and is referred to as a "Modulus-16" counter (or

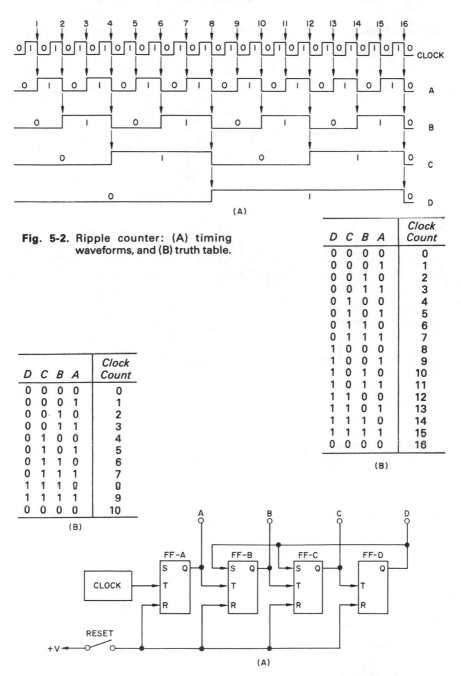

(B)

Fig. 5-2. Ripple counter: (A) timing waveforms, and (B) truth table.

D	C	B	A	Clock Count
0	0	0	0	0
0	0	0	1	1
0	0	1	0	2
0	0	1	1	3
0	1	0	0	4
0	1	0	1	5
0	1	1	0	6
0	1	1	1	7
1	0	0	0	8
1	0	0	1	9
1	0	1	0	10
1	0	1	1	11
1	1	0	0	12
1	1	0	1	13
1	1	1	0	14
1	1	1	1	15
0	0	0	0	16

(B)

D	C	B	A	Clock Count
0	0	0	0	0
0	0	0	1	1
0	0	1	0	2
0	0	1	1	3
0	1	0	0	4
0	1	0	1	5
0	1	1	0	6
0	1	1	1	7
1	1	1	0	0
1	1	1	1	9
0	0	0	0	10

(B)

Fig. 5-3. A simple decade counter: (A) logic circuit, and (B) truth table.

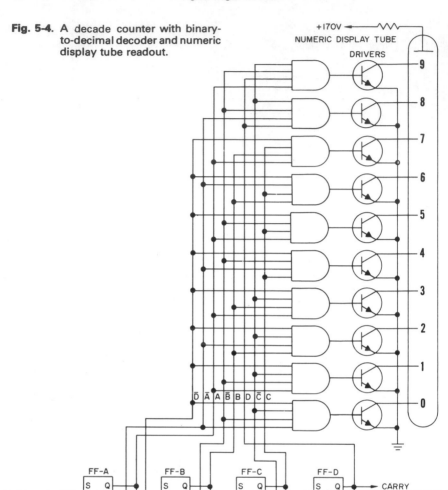

Fig. 5-4. A decade counter with binary-
to-decimal decoder and numeric
display tube readout.

"Mod-16" for short). The decade counter would be referred to as a "Mod-10"
counter. Some manufacturers use the term *modulo*.

PARALLEL COUNTERS

The ripple counter, although simple in operation and amount of components, has a speed limitation. This is due to the fact that each flip-flop is
triggered by the previous flip-flop. Each flip-flop has an inherent time delay

Fig. 5-5. A decimal counter capable of counting to 999.

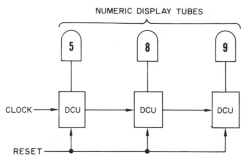

NUMERIC DISPLAY TUBES

which is additive for the number of flip-flops used. The total time required for a complete count is referred to as the *settling time*. This limitation can be overcome through the use of a *parallel* or *synchronous* type counter, where all flip-flops are triggered by the clock pulse and all switch simultaneously.

A typical Mod-8 parallel binary counter is shown in Fig. 5-6. The counter employs J-K flip-flops. When both the J and K inputs are at the 1 level, the flip-flops are enabled, and the counter will count. The incoming pulse is inverted and fed to FF-A. The non-inverted pulse is fed to FF-B and FF-C. FF-A will thus change states on each positive-going pulse (negative-going at the output of the inverter). NAND gate A's output will go negative (lo) each time FF-A's Q = 1 and a clock pulse occurs. FF-B will switch with every other clock pulse. NAND gate B's output will go negative (lo) when FF-A's and FF-B's Q = 1 and a clock pulse occurs. FF-C will thus switch with every fourth clock pulse. The timing waveforms and truth table for the counter are shown in Fig. 5-7.

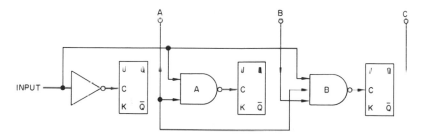

Fig. 5-6. A Mod-8 parallel binary counter. Although not shown, all J and K inputs must be at the one level to enable the flip-flops.

A timing problem that can occur in parallel counters is the *race problem*. The problem occurs from slight variations in the switching time of the gate and flip-flop inputs in a parallel counter. Since all flip-flops must switch simultaneously in a parallel counter, this problem can lead to signal errors. There are a number of ways of coping with the problem. The most widely used method is through the use of master-slave-type, J-K flip-flops (discussed in the previous chapter). This type of flip-flop operates so that it changes state only after the

Digital Logic Circuits

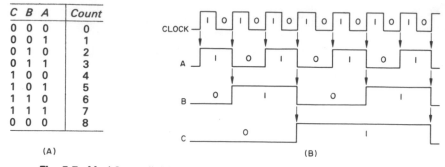

C	B	A	Count
0	0	0	0
0	0	1	1
0	1	0	2
0	1	1	3
1	0	0	4
1	0	1	5
1	1	0	6
1	1	1	7
0	0	0	8

(A)   (B)

Fig. 5-7. Mod-8 parallel binary counter: (A) truth table, and (B) timing
waveforms.

clock pulse has gone low. This prevents any overlapping of pulses and eliminates
the race problem in parallel counters.

A RING COUNTER

If the Q and $\overline{Q}$ outputs of the last flip-flop of a serial counter are fed back to
the J and K inputs of the first flip-flop, we have a *ring counter* (as shown in Fig.
5-8). Note that each J-K flip-flop has direct-set and direct-reset inputs in addition
to the J and K inputs. If all the flip-flops are reset to 0, then if FF-A is set to
Q = 1, and the clock is allowed to run, the 1 will be shifted from FF-A to FF-B
to FF-C to FF-D and then back to FF-A. The 1 will continue to be recirculated through the counter as long as the clock runs.

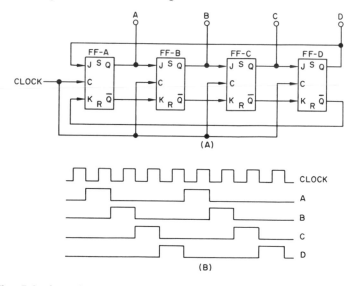

Fig. 5-8. A typical ring counter: (A) logic circuit, and (B) timing
waveforms.

If all the flip-flops are reset to 0, then there will not be any 1 level to circulate, and all the flip-flops will remain at 0. If more than one flip-flop is set high, the word will continue to circulate through the counter in the same manner as the single 1.

A SHIFT COUNTER-DECODER

The feedback method employed in the ring counter can be used in constructing shift-type counters. A decade-shift-type counter and decoder are shown in Fig. 5-9. This decade counter is actually a Mod-5 (FF-A through FF-E) counter followed by a Mod-2 counter (FF-F). Notice that FF-F is set high

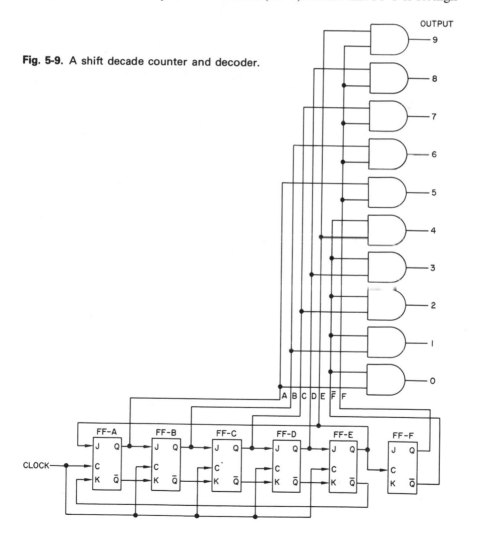

Fig. 5-9. A shift decade counter and decoder.

on counts 5 through 9 and set low on counts 0 through 4. Therefore, only 2-input AND gates are required to decode this decade counter, as compared to the decade counter shown previously in Fig. 5-4, which required 4-input gates.

If the feedback from the last flip-flop to the first flip-flop is inverted, the shift counter is called a *Johnson counter,* as shown in Fig. 5-10. The circuit will divide the clock frequency by 6. If all the flip-flops are set low and then the clock is allowed to run, FF-A will set on the first clock pulse because FF-C's $\overline{Q}$ output is high and fed back to the J input of FF-A. On the second pulse, FF-A will remain high since FF-C's $\overline{Q}$ is still high, and FF-B will switch, since FF-A's $Q = 1$, and therefore FF-B's $J = 1$. On the third clock pulse, FF-C will set high since FF-B's $Q = 1$. On the fourth clock pulse, FF-A will reset since its K input is high. On the fifth pulse, FF-B will reset since its K input is now high, and on the sixth clock pulse, FF-C will reset since its K input is high. Each flip-flop will produce one pulse for each six clock pulses.

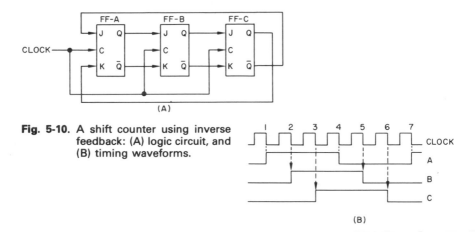

Fig. 5-10. A shift counter using inverse feedback: (A) logic circuit, and (B) timing waveforms.

AN UP-DOWN COUNTER

All of the counters we have examined so far have been up counters. Sometimes we require a down counter. Actually all of the counters can also be used as down counters by utilizing the $\overline{Q}$ outputs instead of the Q outputs. For example, the ripple counter illustrated in Fig. 5-1, as shown in the truth table of Table 5-1, counts up on the A, B, C, D outputs and down on the $\overline{A}$, $\overline{B}$, $\overline{C}$, $\overline{D}$ outputs.

A down counter can be made by connecting the $\overline{Q}$ outputs of each flip-flop to the T input of the following flip-flop. Such an arrangement is shown in Fig. 5-11. FF-A operates as before; FF-B switches when FF-A's $\overline{Q}$ goes low; and so on.

An up-down counter can be made as shown in Fig. 5-12. When the count-up input is 1, gates A and D will be enabled, thus allowing the Q output of

Table 5-1. Truth Table for 4-Bit Ripple Count Q and Q̄ Outputs.

Q̄ Outputs					Q Outputs				
Clock Count	D̄	C̄	B̄	Ā	D	C	B	A	Clock Count
15	1	1	1	1	0	0	0	0	0
14	1	1	1	0	0	0	0	1	1
13	1	1	0	1	0	0	1	0	2
12	1	1	0	0	0	0	1	1	3
11	1	0	1	1	0	1	0	0	4
10	1	0	1	0	0	1	0	1	5
9	1	0	0	1	0	1	1	0	6
8	1	0	0	0	0	1	1	1	7
7	0	1	1	1	1	0	0	0	8
6	0	1	1	0	1	0	0	1	9
5	0	1	0	1	1	0	1	0	10
4	0	1	0	0	1	0	1	1	11
3	0	0	1	1	1	1	0	0	12
2	0	0	1	0	1	1	0	1	13
1	0	0	0	1	1	1	1	0	14
0	0	0	0	0	1	1	1	1	15
15	1	1	1	1	0	0	0	0	16

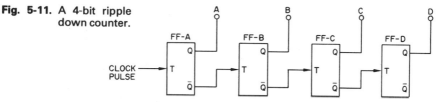

Fig. 5-11. A 4-bit ripple down counter.

each flip-flop to pass its pulse to the following flip-flop, causing the counter to count up in the same manner as previously described. When the count-down input is 1, gates B and E will be enabled, thus allowing the Q̄ outputs to pass their pulse to the following flip-flops and causing the counter to count down in same manner as the circuit in Fig. 5-11.

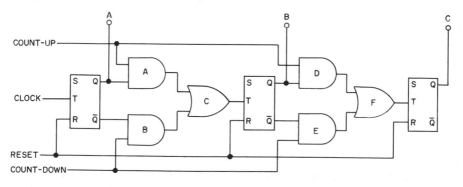

Fig. 5-12. A 3-bit up-down ripple counter.

REGISTERS

Registers are employed to store temporarily binary information such as the data at the output of an encoder, at the input to a decoder, and the input and output of a digital computer. Further, registers can perform the operations of complementing, mulitplication, and division.

A register is a group of flip-flops used to store a binary word. One flip-flop is required for each bit to be stored. The flip-flops are wired so that the binary word can be shifted (entered) into and out of the register. The circuit is generally referred to as a *shift register*. When the word is shifted into the register one bit at a time, the register is called a *serial shift register*. When all the bits are entered simultaneously, the register is called a *parallel shift register*.

A shift register may also be used to convert a serial word to a parallel word, or vice versa, on command. Further, it is possible to enter a word into a register at one clock frequency and shift it out at a different clock frequency. Finally, after the data has been removed from a register, the register can be erased in anticipation of receiving new data. This process is called *clearing* the register.

The Serial Shift Register

A simple 2-bit serial shift register is shown in Fig. 5-13. Assume both flip-flops have been reset to $Q = 0$ and the flip-flops go through their transition

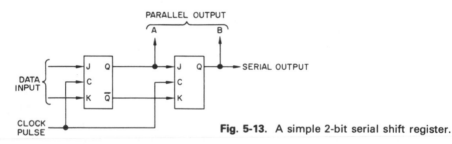

Fig. 5-13. A simple 2-bit serial shift register.

on the negative-going part of the clock pulse. If both J and K have a 0 input, neither flip-flop will switch on the clock pulses. If the J input of FF-A $= 1$, then on the next clock pulse FF-A will set $Q = 1$ and $\overline{Q} = 0$. Now if FF-A's J input goes to 0 and $K = 1$, FF-A will reset to $Q = 0$ on the next clock pulse, and FF-B will switch to $Q = 1$. In other words, the 1 has been shifted from FF-A to FF-B. On the next clock pulse FF-B will reset to $Q = 0$ and the 1 will have been shifted out of the register.

Restating the operation another way, if $J = 1$ and $K = 0$, FF-A will set to 1 on the next clock pulse. If $J = 0$ and $K = 1$, FF-A will reset to 0 on the next clock pulse. If FF-A's $Q = 1$, FF-B will set to 1 on the next clock pulse. If FF-A's $Q = 0$, FF-B will reset to 0 on the next clock pulse.

Figure 5-14 shows a 5-bit serial shift register. It is capable of storing a 5-bit word. Suppose we wish to store the 5-bit word 10110 in the register. The register

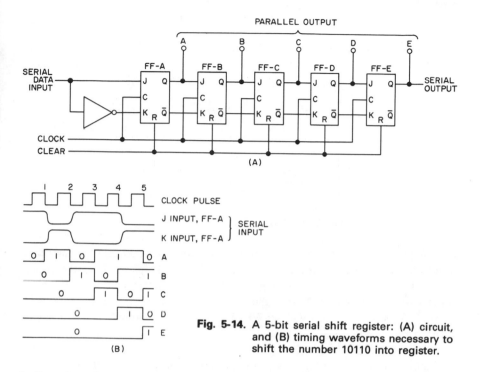

Fig. 5-14. A 5-bit serial shift register: (A) circuit, and (B) timing waveforms necessary to shift the number 10110 into register.

is first cleared by a pulse to the reset inputs. The word is then fed into the data input. The first clock pulse will shift the first bit (1) into FF-A. The second clock pulse will shift the second bit (0) into FF-A, while transferring the first bit (1) to FF-B. The third clock pulse will shift the third bit (1) into FF-A, shift the second bit (0) to FF-B and shift the first bit (1) to FF-C. The fourth clock pulse shifts the fourth bit (1) into FF-A, while shifting the prior bits to FF-B, FF-C and FF-D. The fifth clock pulse shifts the fifth bit (0) into FF-A, while shifting the prior bits to FF-B, FF-C, FF-D, and FF-E, in succession. At this point, if the clock input is stopped, the word will be stored in the register. If the clock is not stopped the word will be shifted out of the register.

The Parallel Shift Register

The serial shift register is simple, but slow, since one clock pulse is required for each bit to be shifted into the register. The parallel shift register permits a word, in parallel form, to be shifted into the register with a single clock pulse (after the register has been cleared).

A three-bit parallel shift register is shown in Fig. 5-15. When the AND gates A, B, and C are enabled by the shift input = 1, the levels at the 1, 2, and 4 inputs will be shifted into the register on the next clock pulse. When the shift and reset pulses are removed, the word will remain in the register, even though the clock continues to run, since both the J and K inputs will be 0.

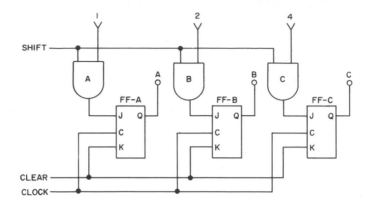

Fig. 5-15. A 3-bit parallel shift register.

A parallel shift register that does not have to be cleared prior to entering a word is shown in Fig. 5-16. This is twice as fast as the previous shift register, where one clock pulse was first required to clear the register. A 1 level at the 1 input will set FF-A Q = 1, regardless of its previous level, on the next clock pulse. Likewise, a 0 level will set FF-A to Q = 0, regardless of the previous level in the flip-flop.

Recall that in Chap. 2 the 1's complement was often used in binary arithmetic. Shift registers can be used for complementing a word stored in the register. If the J and K inputs of each flip-flop are enabled for one clock pulse, then each flip-flop will change state, and the output of the register will now be the complement of the word previously stored in the register.

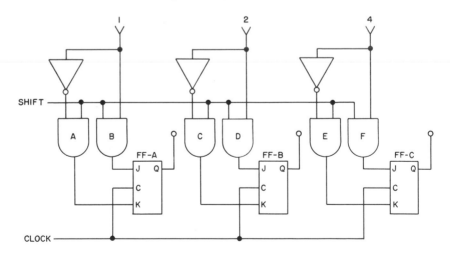

Fig. 5-16. A parallel shift register not requiring clearing before entering a word.

Left-Right Shift Register

The data stored in a register may be shifted to the left or to the right within the register. Shifting data to the left is equivalent to multiplication by 2 for each place shifted. For example, if a register has the decimal number 3 (0011) stored in it and the bits are shifted one place to the left, the binary number becomes 0110, which is 6 in decimal. If the bits are shifted one more place to the left, the binary number becomes 01100, which is 12 in decimal. Conversely, if a binary number is shifted to the right, it is equivalent to dividing its decimal equivalent by 2. Naturally, the register must have sufficient size to accomplish the shifting without shifting any bits out of the register, or errors will occur.

Figure 5-17 illustrates the basic operation of a shift-left and a shift-right register. The word in the shift-left register will be shifted left by one place each time the clock pulse occurs and the shift-left input is high.

INTEGRATED CIRCUIT COUNTERS AND REGISTERS

Through the technology of integrated circuits, it is possible to have entire counters and shift registers made in functional blocks. A wide variety is available. A few examples follow.

A decade counter providing an 8421 BCD output (Q_1 through Q_4) is shown

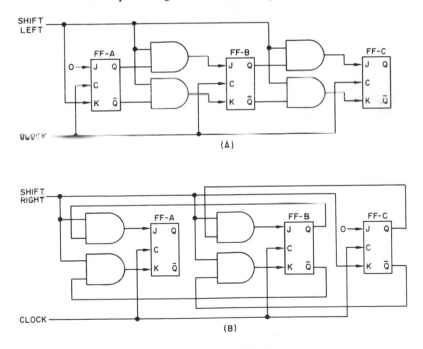

Fig. 5-17. (A) Shift-left register, and (B) shift-right register.

in Fig. 5-18. The serial input is fed in at the CP terminal. The counter employs R-S flip-flops, incorporating direct-set (S_D) and direct-reset (C_D) inputs. The unit is available in standard 14-pin ceramic flat pack or DIP plastic packages.

MC938F · MC838F, P

This monolithic ripple counter is designed to operate at ±20% of the nominal 5.0 volt power supply voltage and is guaranteed to 15 MHz. It has standard MDTL inputs, and uses active pull-up devices in the outputs to increase capacitive drive capabilities. The outputs correspond to a standard 8-4-2-1 BCD with individual direct sets and a common direct clear available to preset the counter to any desired condition. Typical noise margin is 1.0 volt. Typical input fall time required to toggle is 5 μs measured from 2.5 volts to 0.50 volts.

Input Loading Factor
S_D = 1.5
C_D = 5
CP = 1

Output Loading Factor = 8

Total Power Dissipation = 150 mW typ/pkg

Maximum Counting Frequency = 30 MHz typ

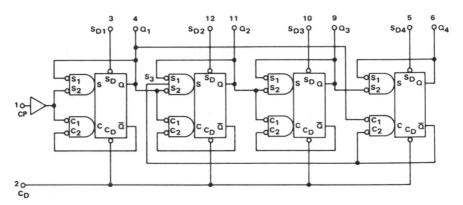

COUNTER SEQUENCE

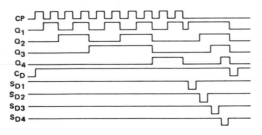

DECODING LOGIC

0	$\bar{Q}_1$	$\bar{Q}_2$	$\bar{Q}_3$	$\bar{Q}_4$
1	Q1	$\bar{Q}_2$	$\bar{Q}_3$	$\bar{Q}_4$
2	$\bar{Q}_1$	Q_2	$\bar{Q}_3$	
3	Q_1	Q_2	$\bar{Q}_3$	
4	$\bar{Q}_1$	$\bar{Q}_2$	Q_3	
5	Q_1	$\bar{Q}_2$	Q_3	
6	$\bar{Q}_1$	Q_2	Q_3	
7	Q_1	Q_2	Q_3	
8	$\bar{Q}_1$			Q_4
9	Q_1			Q_4

Fig. 5-18. An IC decade counter.
(Courtesy Motorola)

Fig. 5-19. 5-bit shift register SN7496.
(Courtesy Texas Instruments)

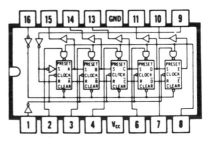

PROPAGATION DELAY   25 NS
POWER DISSIPATION   245 MW

A 5-bit shift register is shown in Fig. 5-19. It is a type-7496 and consists of five R-S master-slave flip-flops connected as a shift register to perform parallel-to-serial or serial-to-parallel conversion of binary data. The flip-flops may be set independently by applying a pulse to the flip-flop's "preset" input. All the flip-flops are cleared simultaneously and clocked simultaneously. Each output has a buffer amplifier to isolate the flip-flops from their loads.

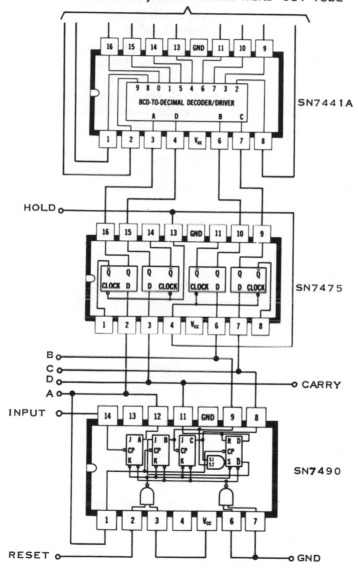

Fig. 5-20. Storage and readout of BCD data from high-speed counter. (*Courtesy Texas Instruments*)

Figure 5-20 shows how only three ICs can be used to provide all the digital logic necessary to operate a cold-cathode, gas-filled numeric display tube. The serial input is fed into pin 14 of the SN7490 decade counter. The decade counter is composed of three J-K and one RST flip-flops interconnected so that the counter has only ten possible states. The 8421 BCD output of the counter (A, B, C, and D) appears on terminals 8, 9, 11, and 12. A carry pulse can be taken from pin 11 to drive a second decade counter-decoder-readout circuit. The BCD output is fed to an SN7475 quadruple bi-stable latch, composed of four storage flip-flops, each having Q and $\overline{Q}$ outputs. The binary number present at the data (D) inputs are transferred to the Q outputs if the "hold" input is high. When the hold input goes low, the binary number that was present at the data inputs at the time this transition occurred is retained at the Q outputs. The SN7441A IC receives the binary number from the latch and decodes it into a decimal number so that the output for the particular digit is enabled. Each decimal output has a driver capable of operating the cold-cathode numeric display tube.

Review Questions

1. A ripple counter circuit uses six flip-flops. How high can it count?
2. What is a DCU?
3. If we wish to count up to 999,999, how many DCU's will we need?
4. Show a truth table for the counter shown below. How high will it count?

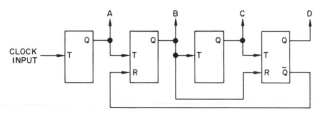

5. Show a truth table for the counter shown below. How high will it count?

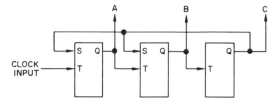

6. Why is a parallel counter faster than a serial counter?
7. Make a truth table for the up-down counter shown in Fig. 5-12.
8. Describe the difference between the methods of entering data into serial and parallel shift registers.
9. How can a shift register be used to obtain the 1's complement of a number?
10. How can a shift register multiply a number by 4?

Chapter 6

Arithmetic Circuits

In the preceding chapters, we studied circuits for inputting and outputting data from a digital system such as a computer. In addition, we investigated circuits for counting and storing binary data. In this chapter, we will examine some typical circuits employed in performing arithmetic functions such as addition, subtraction, multiplication and division.

THE EXCLUSIVE-OR GATE

A standard 2-input OR gate is enabled if one or both inputs = 1. The exclusive-OR gate, shown in Fig. 6-1 is enabled if one or the other, but not both, inputs = 1. If the A input = 1 and the B input = 0, then AND gate A is enabled, in turn enabling the OR gate and producing a 1 output. However, if A and B = 1, or if A and B = 0, both AND gates are disabled, in turn disabling the OR gate and yielding a 0 output.

The exclusive-OR gate derives its name from the fact that it compares the two input levels. If they are the same it produces a 0 output. If they are different it produces a 1 output. The circuit is sometimes referred to as a *comparator*. The circuit may be designed in other ways than that shown in Fig. 6-1 (refer to Fig. 3-8); however, an exclusive-OR symbol (refer to Fig. 6-1(c)) is often used to represent the exclusive-OR function.

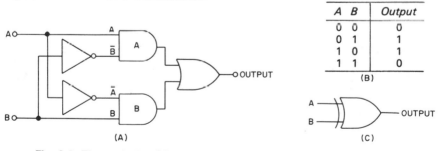

A	B	Output
0	0	0
0	1	1
1	0	1
1	1	0

(B)

(A)

(C)

Fig. 6-1 The exclusive-OR gate: (A) logic circuit, (B) truth table, and (C) logic symbol.

PARITY-BIT ERROR CHECKING

In Chap. 2, it was pointed out that a popular method of error checking was to add a parity-bit to a binary number. A 1 is added if the binary number contains an

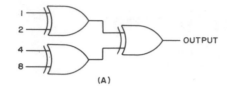

8	4	2	1	Output
0	0	0	0	0
0	0	0	1	1
0	0	1	0	1
0	0	1	1	0
0	1	0	0	1
0	1	0	1	0
0	1	1	0	0
0	1	1	1	1
1	0	0	0	1
1	0	0	1	0
1	0	1	0	0
1	0	1	1	1
1	1	0	0	0
1	1	0	1	1
1	1	1	0	1
1	1	1	1	0

(A)

Fig. 6-2. A parity-bit generator employing three exclusive-OR gates: (A) logic circuit, and (B) truth table.

(B)

odd amount of 1's; conversely, a 0 is added if the binary number contains an even amount of 1's. A circuit for generating the parity-bit is shown in Fig. 6-2. The circuit consists of three exclusive-OR gates having the standard 8421 BCD code inputs. The truth table demonstrates that the circuit generates a 0 when the binary number has an even number of 1's, and it generates a 1 when the binary number has an odd number of 1's.

The parity-bit may be compared to the number of 1's in the binary number at any point in the system, to check for errors. For example, when a binary number is transmitted from one computer to another, the receiving computer would check for parity. If it is found that the number is in error, then the receiving computer would request the sending computer to retransmit the data.

EQUALITY CHECKING

Frequently, in a digital system, it is necessary to compare two words and determine whether they are equal. The circuit shown in Fig. 6-3 is an equality detector and can perform this function. Each bit of a 4-bit word is compared by an exclusive-OR gate. The outputs from each comparator are fed to an AND gate to determine if the two words, as a whole, are equal. If the two words are equal, the AND gate will have a 1 output. If they are not equal, the output will be a 0.

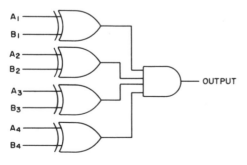

OUTPUT

Fig. 6-3. A 4-bit equality detector using exclusive-OR gate.

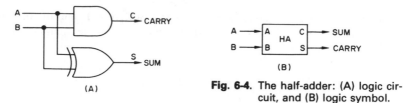

Fig. 6-4. The half-adder: (A) logic circuit, and (B) logic symbol.

THE HALF-ADDER

A circuit which adds two binary digits is shown in Fig. 6-4. It is called a *half-adder* (HA). The sum of two binary numbers appears at the output of the exclusive-OR gate, while the carry appears at the output of the AND gate. There are four possible states for the half-adder. They are as follows:

Input		Output	
A	B	Sum	Carry
0	0	0	0
0	1	1	0
1	0	1	0
1	1	0	1

THE FULL-ADDER

The full-adder logic circuit (FA) can sum two binary digits plus the carry from a previous half-adder or full-adder. A typical full-adder logic circuit is shown in Fig. 6-5. If the carry-in input = 0, the circuit functions just like a half-adder. For example, if the inputs to the full-adder are A = 1, B = 1, and carry-in = 0, then the first half-adder outputs are C = 1 and S = 1. The inputs to the second half-adder are A = 0 and B(C_{in}) = 0, and the outputs are C = 0 and S = 0. Hence, the inputs to the OR gate are 1 and 0, yielding a 1 output. The net result is that the outputs of the full-adder are C = 1 and S = 0.

As another example, if the inputs to the full-adder are A = 1, B = 1, and C_{in} = 1, the output of the first half-adder will be C = 1 and S = 0. The

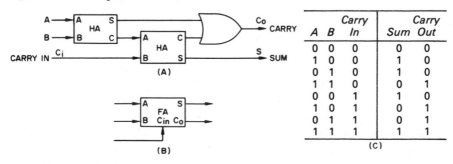

	Carry		Carry	
A	B	In	Sum	Out
0	0	0	0	0
1	0	0	1	0
0	1	0	1	0
1	1	0	0	1
0	0	1	1	0
1	0	1	0	1
0	1	1	0	1
1	1	1	1	1

(C)

Fig. 6-5. The full-adder: (A) logic circuit, (B) logic symbol, and (C) truth table.

inputs to the second half-adder are A = 0 and $B(C_{in})$ = 1, and the outputs are
C = 0 and S = 1. The OR gate will produce a 1 output (carry), and the S
output will also be 1.

THE PARALLEL ADDER

Thus far, only single-digit binary numbers have been added. The addition
of multi-digit binary number may be performed in a parallel or serial adder
circuit. In the parallel adder, all digits are added simultaneously. An adder is
required for each set of digits to be added. Hence, in the circuit shown in Fig.
6-6, three adders (two FA's and one HA) are required to add two 3-bit binary
numbers simultaneously.

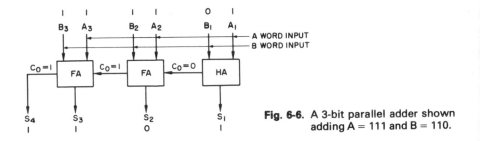

Fig. 6-6. A 3-bit parallel adder shown adding A = 111 and B = 110.

The column on the right (least significant digit) may be added using a
half-adder since there will not be any previous carry. The following adders,
however, must be full-adders since there may be a carry from the previous
adders.

As an example, if we were to add 111 (word A) and 110 (word B), the
operation would be as follows:

$$A = 111$$
$$B = +110$$
$$\overline{\text{sum} = 1101}$$

The half-adder will produce S_1 = 1 with no carry. The first full-adder will
have an output of S_2 = 0 with a carry of 1. The second full-adder will have an
output of S_3 = 1 and also a carry = 1.

The parallel adder circuit is employed in most high-speed computer systems
since the propagation time of the parallel adder is much less than that of the
serial-type adder. In most high-speed computers, sensing is incorporated into the
adder logic to check the state of the next lower-order adder stages from which the
carry would come and therefore can anticipate the carry. This reduces the carry
propagation time by two or three times. This sensing is called a *look-
ahead-for-carry* feature.

Fig. 6-7. A serial adder unit.

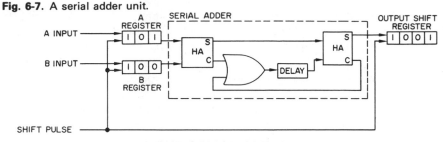

THE SERIAL ADDER

Serial addition is performed using two half-adders and entering each binary number serially with the least significant digit (LSD) entered first and the most significant digit (MSD) last. A basic serial adder, shown in Fig. 6-7, has two half-adders, an OR gate, and a delay device.

The serial adder operates step-by-step, adding the LSD first and progressing to the more significant digits in turn. For example, in adding the binary numbers 101 and 100, the LSD 1 and 0 are added first, yielding sum = 1, the next more significant digits are added to obtain sum = 0, and finally the MSD 1 and 1 are added to yield a sum = 1 and carry = 1.

The operation of the circuits shown in Fig. 6-7 is as follows: The binary numbers are read into the A and B shift registers. Both registers are then shifted, one digit at a time, upon the receipt of a shift pulse. The adder performs the addition, one bit at a time, shifting its sum into the output register. The first half-adder provides a sum and possible carry, which is applied to the second half-adder. The delay device delays the carry pulse one repetition period of the shift pulse so that the second half-adder adds the sum output of the digit applied to the serial adder input to a carry from the previous digit. The final sum is generated one repetition period after the final shift pulse occurs. The last possible carry travels through the delay device during this time and becomes a sum output.

A more economical serial adder is shown in Fig. 6-8. It employs one shift register to store both of the input binary numbers and the sum output of the serial adder. This register is called an *accumulator* register, while the register accepting the number from the outside source is called the *incident* register. This type of adder can add more than two numbers by accumulating the sum of successive additions.

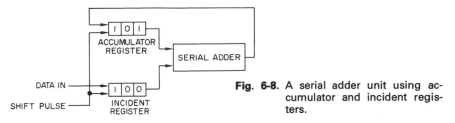

Fig. 6-8. A serial adder unit using accumulator and incident registers.

The operation of this serial adder is as follows: All registers are first cleared; the first number is read into the incident register and then shifted through the adder into the accumulator register as the second number is read into the incident register. Since the accumulator register was first set to 0000, the number is not affected by the adder. On the following shift pulses, the accumulator and incident register contents are added and their sum stored in the accumulator register. If a new number is read into the incident register, it can be added to the sum already in the accumulator register.

AN 8421 BCD ADDER

When a BCD code (e.g., 8421) is used, recall (Chap. 2) that each decimal digit is encoded to a 4-bit equivalent. For example, 6243 would be encoded to 0110 0010 0100 0011. Further, there are no 4-bit groups above 1001, since only 10 of the possible 16 4-bit groups are used. It is therefore necessary, when the sum is greater than 9, to skip six groups to restore the number to its correct sum. For example, if we add decimals 7 and 6, we get a BCD number which is not used, and then 6 must be added to the number to obtain the correct BCD number for the sum.

$$
\begin{array}{ll}
0111 & 7 \\
+0110 & +6 \\
\hline
1101 = & 13 \quad \text{in binary} \\
+0110 & +6 \\
\hline
0001\ 0011 = & 13 \quad \text{in BCD code}
\end{array}
$$

If the sum is less than 9, it is correct for the BCD code without the addition of 6.

A BCD adder to perform the necessary operations is shown in Fig. 6-9. If we were adding 7 (0111) and 6 (0110), the operation of the adder would be as follows: The top row of adders would add the two numbers to yield an output of 01101; AND gate A would be enabled, in turn enabling the OR gate and feeding a 1 into the A inputs of the first two full-adders of the bottom row. The output of the BCD adder circuit would therefore be 0011 with a carry of 1 to the next adder. This is equal to 13 in BCD code. AND gates A and B together with the OR gate sense any number greater than 1001 (9), in turn causing the bottom row of full-adders to add 0110 (6) to the output. When the BCD number is 9 or less, 0000 is added to the number. By cascading BCD adders we can add numbers of any size.

SUBTRACTORS

A parallel adder can be made into a subtractor by feeding the 1's complement of one of the numbers into the adder, adding the numbers and doing an

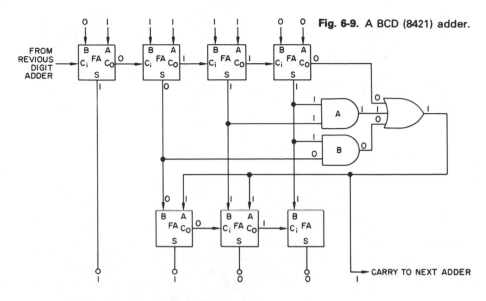

Fig. 6-9. A BCD (8421) adder.

end-around-carry, as was described in Chap. 2. Such a circuit is shown in Fig. 6-10 and is referred to as a *parallel binary subtractor*.

The 1's complement of a number is usually taken from the $\overline{Q}$ outputs of the parallel shift register feeding the adder, or in some cases, as shown, inverters are employed. If we wish to subtract 3 (011) from 7 (111), the circuit functions as follows: 011 is complemented to 100. The adder then adds 111 and 100. FA-1 and FA-2 initially produce S = 1 and C_0 = 1, and FA-3 produces S = 0 and C_0 = 1. The FA-3 carry is fed back to FA-1, which causes it to change to S = 0 and C_0 = 1. FA-2 changes to S = 0 and C_0 = 1, which cause FA-3 to change to S = 1. The output is thus 100 (4).

A half-subtractor (HS), similar to the half-adder, can also be constructed. Such a circuit is shown in Fig. 6-11. This circuit subtracts one binary digit from another. When X = 0 and Y = 0, the subtractor outputs are borrow = 0 and difference = 0 (0 − 0 = 0). When X = 0 and Y = 1, the outputs are borrow = 1 and difference = 1 (0 − 1 = 1 with a borrow = 1).

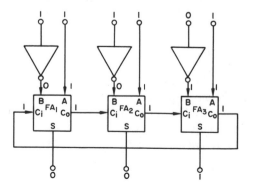

Fig. 6-10. A parallel binary subtractor unit.

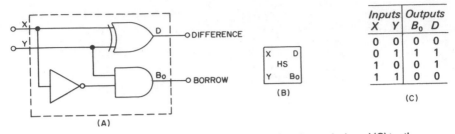

Inputs		Outputs	
X	Y	B_0	D
0	0	0	0
0	1	1	1
1	0	0	1
1	1	0	0

(C)

Fig. 6-11. A half-subtractor: (A) logic circuit, (B) logic symbol, and (C) truth table.

When X = 1 and Y = 0, the borrow = 0 and difference = 1 (1 − 0 = 1 with a borrow = 0). When X = 1 and Y = 1, the outputs are borrow = 0 and difference = 0 (1 − 1 = 0 with a borrow = 0). These outputs are consistent with the rules of binary subtraction established in Chap. 2.

A full-subtractor (FS) can find the difference between two binary digits and the borrow from a previous half-subtractor or full-subtractor. A full-subtractor is shown in Fig. 6-12. Half- and full-subtractors can be cascaded to find the difference between multi-digit binary numbers in the same manner as full-adders.

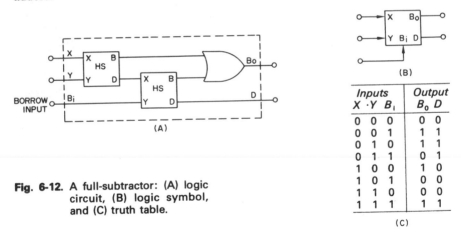

Inputs			Output	
X	Y	B_i	B_0	D
0	0	0	0	0
0	0	1	1	1
0	1	0	1	1
0	1	1	0	1
1	0	0	1	0
1	0	1	0	0
1	1	0	0	0
1	1	1	1	1

(C)

Fig. 6-12. A full-subtractor: (A) logic circuit, (B) logic symbol, and (C) truth table.

ADDER UNIT ORGANIZATION

Thus far, the adders and subtractors we have examined deal in whole numbers. In a working computer or calculator, we deal in fractions as well as whole numbers. This is accomplished by inserting a decimal point by either of two methods, the fixed-point, or the floating-point. An arithmetic unit using the fixed-point method is called a *binary fractional machine,* since all numbers will be represented as a fraction.

Another requirement of the adder unit is that it must have a way of indicating whether a number is positive or negative. It should also be able to perform arithmetic operations between signed numbers. Lastly, the adder unit must be able to sense and indicate when the contents of a register is exceeded. When this occurs, the adder unit should indicate that an "overflow" condition exists and that an error exists in the computations.

The following sections describe how these adder requirements are implemented. The presentation is reduced to its basic essentials and illustrates only one possible method.

FIXED-POINT ARITHMETIC

In a fixed-point adder all numbers are represented as fractions with the decimal point on the left side of the MSB (most significant bit). With this procedure we can work with both fractions and whole numbers, making full use of the registers and enabling easy detection of overflow conditions. For example, the decimal number 28.125 in binary would be 11100.001. If we shift the decimal point five places to the left, the number becomes .11100001 $\times$ 2^5. The 2^5 tells us that the decimal point has been shifted five places to the left of the MSD. It would require an 8-bit register to store this number.

To indicate positive and negative numbers we use one bit to the left of the decimal point. A 0 represents a negative number while a 1 represents a positive number. For example, 0.11100001 $\times$ 2^5 = -28.125. Therefore, it would require a 9-bit register to store this number.

Further, we may need to use the 1's or 2's complement of the number. The 1's complement of -28.125 is 1.00011110 $\times$ 2^5 and the 2's complement is 1.00011111 $\times$ 2^5. Notice that the sign bit does not change when a number is complemented.

This process of shifting the decimal point is called *scaling,* and is done when the number is entered into the computer. When the result is outputted from the computer, the binary number's decimal must be shifted to the right by the proper number of places. In this case, it would be shifted five places to the right.

The binary adder must be able to perform four possible cases of addition. They are:

1. Addition of two positive numbers
2. Addition of a positive number and a smaller negative number
3. Addition of a positive number and a larger negative number
4. Addition of two negative numbers

When adding two like sign numbers (cases 1 and 4) there is the possibility of overflow, and this must be detected to indicate an error condition. When an overflow occurs, this will cause the sign digit to change. An overflow cannot occur when adding numbers with opposite signs.

THE SERIAL ADDER UNIT

An adder unit to perform the addition of signed mixed numbers with overflow detection is shown in Figs. 6-13, 6-14, and 6-15.

The adder unit utilizes the serial adder circuit shown previously in Fig. 6-8. To this circuit we add first an overflow-detecting circuit, such as is shown in Fig. 6-13. The exclusive-OR gate compares the first digit in the accumulator and incident registers. If they are the same, the exclusive-OR gate output = 0, and the inverter changes the output to 1 to set the J-K flip-flop when the first clock pulse occurs. If the sign-digits are not the same, the J-K flip-flop will not set.

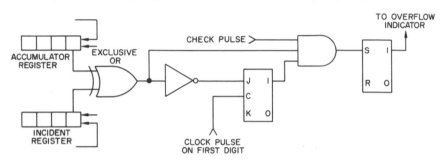

Fig. 6-13. Overflow detection circuit added to serial adder unit.

The adder will now go through the addition cycle with the sum appearing in the accumulator register. A check-overflow pulse now is fed to the 3-input AND gate while the exclusive-OR gate again compares the first digits in the two registers. If the exclusive-OR gate output = 1 and the J-K flip-flop was previously set, then the R-S flip-flop will = 1 to indicate an overflow error.

The next step is incorporating a 2's complementing circuit. Actually, a 1's complementing circuit could be employed, but this would require two cycles for each addition to accomplish the addition and end-around-carry. The 2's complement addition, as shown in Chap. 2, does not require this end-around-carry and is therefore the most widely used method in serial systems.

A circuit for 2's complementing of a serial adder is shown in Fig. 6-14. The circuit complements every bit, except the sign bit and the first 1, beginning with

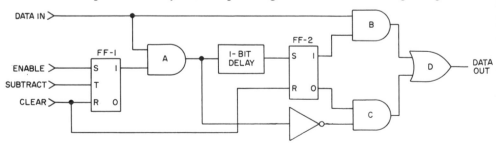

Fig. 6-14. Circuit for 2's complement of a serial adder.

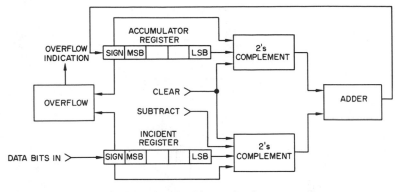

Fig. 6-15. A serial adder and subtractor unit.

the LSB. To begin addition, FF-1 and FF-2 are reset. FF-1 is then set, and AND gate A is enabled for the time of the number digits (but not the sign digit), when a negative sign is detected. The entering data bits pass through AND gate B until after the first 1 occurs. The first 1 is delayed one bit-time and then sets FF-2, disabling AND gate B and enabling AND gate C. The data bits are now routed through the inverter and complemented. Since the first 1 was not complemented, the result is the complement plus 1, or in other words, the 2's complement. FF-1 and FF-2 are reset after the number digits pass through, enabling gate B and disabling gate C. This insures that the sign bit is not changed.

Recall also from Chap. 2, that subtraction of two binary numbers can be accomplished by taking the 2's complement, adding the two numbers, and dropping the carry. Hence, we can use the circuit in Fig. 6-14 to do subtraction by taking the 2's complement of the serial number and adding the two numbers serially in the adder unit. If a number to be subtracted is a negative number, the sensing of the negative (−) sign will disable the 2's complementer, and the number will be added as an ordinary binary number. To accomplish subtraction then, a subtract control pulse is fed to FF-1 to enable it and form the 2's complement.

A serial binary adder unit with overflow detection and capability of adding and subtracting both positive and negative numbers is shown in Fig. 6-15.

THE PARALLEL ADDER UNIT

The parallel adder unit is much faster than the serial adder unit since it requires only one clock pulse to provide the sum of two numbers. However, the parallel adder requires considerably more circuitry than the serial adder to sum numbers to comparable size.

A parallel adder-subtractor unit is shown in Fig. 6-16. Its operation is as follows: The registers are first cleared. The A-word is read into the incident

Digital Logic Circuits

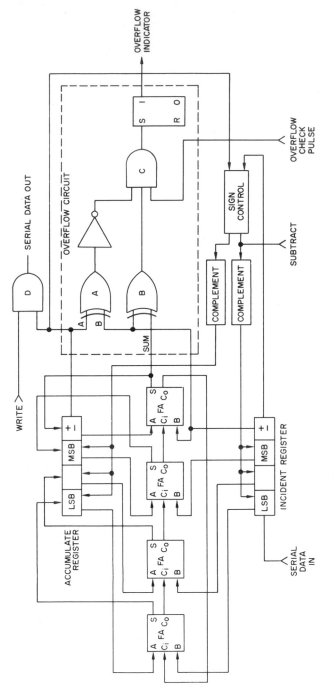

Fig. 6-16. Basic parallel binary adder-subtractor unit.

register serially. For faster operation, an additional *Memory Buffer Register* (MBR) may be incorporated. If an MBR is used, the word is transferred serially from the memory unit to the MBR and then parallel transferred from the MBR to either the accumulator or incident registers. If no MBR is used, the A-word is parallel transferred through the adders to the accumulator register and the B-word is read into the incident register.

On the first add-clock-pulse the sign digits are sensed, and the 2's complement is formed in the respective registers if a negative sign (1) is present. If the numbers are to be subtracted, the incident register word is complemented. On the second clock pulse, the A and B words are parallel transferred through the adders and the sum placed back in the accumulator register. Hence, only two clock pulses are required for parallel addition.

An overflow can occur only when adding two positive or two negative numbers. Hence the exclusive-OR gate A and inverter will present a 1 to AND gate C at this time. Further, gate B will have a 1 output if the B-sign and sum-sign bits are different. Now, when an overflow check pulse is fed to AND-gate C it will be enabled, setting the R-S flip-flop and producing an overflow indication.

The sum may be read out of the accumulator register when AND-gate D is enabled by a "write" input. If an MBR is employed, the word in the accumulator register may be read out in parallel form into the MBR and then from the MBR transferred serially into the memory unit.

Review Questions

1. What is the output of an exclusive-OR gate for the four possible inputs?
2. What is the difference between a half-adder and a full-adder?
3. What is the difference between a parallel adder and a serial adder?
4. Which type of adder is employed where speed is important? Explain why.
5. How can a parallel adder be changed to a subtractor?
6. How and why are overflows detected?
7. What is the difference between "fixed-point" and "floating-point"?

Chapter 7

Pulse Sources
and Shapers

All digital systems operate with changes between the 0 and 1 levels with respect to time. These changes, such as those appearing in Fig. 7-1, are referred to as *pulses*. The pulses may be developed when an operator depresses a switch, a card reader senses a hole in a card, or from the internal clock oscillator of the digital system. There are an infinite number of pulse sources. Chapter 9 covers pulses generated by system operator means. In this chapter, we are concerned with the generation and shaping of pulses generated electronically, usually within the digital electronics of the system.

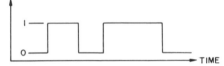

Fig. 7-1. Digital pulses.

When the pulse source is used as the basic timing control for the digital system, it is referred to as the *clock*. As such, it provides a periodic pulse to synchronize two or more circuits which must operate together. In many cases, such as A/D conversion (covered in Chap. 10), the clock frequency is critical and frequency changes may cause errors. Hence, frequency stability of the clock is very important.

MULTIVIBRATORS

In Chap. 4, we examined bistable multivibrators (flip-flops), and observed that they had two stable states. Further, they could switch states when pulsed. If feedback is provided from one transistor to the other and back again, the circuit will change states of its own accord. In other words, it will oscillate back and forth between its two states. The circuit is then called a free-running, or astable, multivibrator. Such a circuit is shown in Fig. 7-2.

The free-running multivibrator may be thought of as two amplifiers, the output of each amplifier being fed to the input of the other amplifier through coupling capacitors (C_1 and C_2). In fact, this circuit is often constructed using two inverters or gates, such as shown in Fig. 7-3.

The operation of the multivibrator is as follows: Initially, one transistor will conduct to saturation because its gain will be slightly greater than the other transistor. Assume Q_2 conducts to saturation; C_1 charges up to $+V_{cc}$ through the

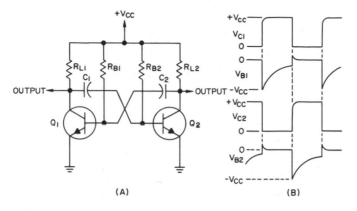

Fig. 7-2. Free-running multivibrator.

base-emitter conduction of Q_2. Since the collector of Q_2 is at essentially ground potential (transistor saturated) C_2 discharges through R_{B1} to V_{cc}, putting a negative potential on the base of Q_1 and cutting it off. Q_2 will remain saturated and Q_1 cut off until C_1 reaches full charge, at which time the base-to-emitter current of Q_2 ceases. Q_2 is no longer forward-biased and it turns off, causing its collector voltage to go to $+ V_{cc}$. Capacitor C_2 will now charge to V_{cc} through the base-to-emitter junction of Q_1 causing Q_1 to conduct to saturation. The collector of Q_1 will now be close to ground potential, causing C_1 to discharge through R_{B2}, holding Q_2 cut off. When C_2 has reached full charge (V_{cc}), Q_1 will again turn off and the process will repeat itself. The waveforms developed are shown in Fig. 7-2(B).

The frequency of oscillation is determined by the time-constant of each capacitor and base resistor through which it discharges. The frequency (f) will be equal to:

$$f = \frac{1}{1.4 \; R_B C}$$

The time (t) for one complete cycle will be equal to:

$$t = 1.4 \; R_B C$$

The base frequency and time of the oscillator in Fig. 7-3 will be:

$$f = \frac{1}{1.4 \times 22 \text{ k} \times 0.001\mu} \qquad = 32.5 \text{ kHz}$$

$$t = 1.4 \times 22 \text{ k} \times 0.001\mu \qquad = 30.8 \text{ μsec}$$

The frequency of the multivibrator may be varied by changing either the values of R_B or C. In Fig. 7-3, the oscillator frequency is varied by changing the value of R_{B2}. The maximum frequency and the time of the oscillator may be calculated as follows:

$$f = \frac{1}{(0.7R_{B1}C) + (0.7R_{B2}C)}$$

$$= \frac{1}{(0.7 \times 22k \times 0.001\mu + (0.7 \times 32 \; k \times 0.001\mu)}$$

$$= 25.4 \; kHz$$

$$t = (0.7R_{B1}C) + (0.7R_{B2}C)$$

$$= 15.6 \; \mu sec + 24 \; \mu sec = 39.4 \; \mu sec$$

The output waveforms of the oscillator in Fig. 7-3 are shown in Fig. 7-4.

The multivibrator oscillator is popular because of its simple and economic circuitry. However, its frequency is affected by changes in temperature and supply voltage. It is therefore used where frequency stability is not a requirement.

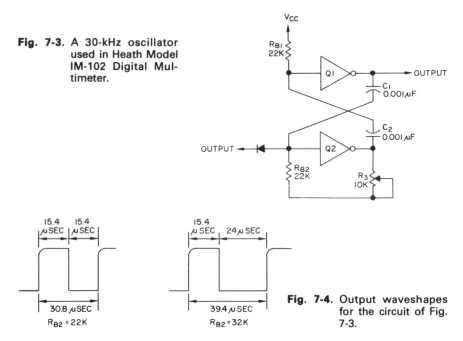

Fig. 7-3. A 30-kHz oscillator used in Heath Model IM-102 Digital Multimeter.

Fig. 7-4. Output waveshapes for the circuit of Fig. 7-3.

The Unijunction Transistor Oscillator

The unijunction oscillator is popular because of its exceptional simplicity. It is basically a relaxation-type oscillator. The unijunction transistor has a special characteristic of negative resistance. The unijunction transistor is shown in Fig. 7-5. It is made of an N-type silicon bar with a P-type diffusion close to the B_1 end of the bar. The connections are called emitter (E), base 1 (B_1), and base 2 (B_2). The resistance of the N-type silicon bar is generally 4–12 kohms.

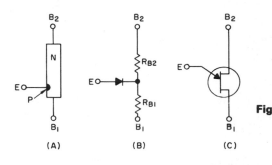

Fig. 7-5. The unijunction transistor:
(A) physical construction,
(B) device characteristics,
and (C) schematic symbol.

The operation of the unijunction transistor as an oscillator can be explained by referring to Fig. 7-6. When voltage is initially applied, a small current flows from B_1 to B_2. C_1 is at zero charge and the voltage at $E = 0$ volts. There is a voltage drop across the R_{B1} resistance (Fig. 7-5) of the silicon bar and hence reverse-biasing of the E-B_1 junction.

Capacitor C_1 now starts to charge to $+V$ through R_1. When the voltage on the capacitor exceeds the voltage drop across the R_{B1} resistance, the junction is forward-biased. Hole carriers now cross over into the R_{B1} region, greatly reducing its resistance. This provides a low-resistance path for C_1 to discharge. When C_1 has discharged, there is no further injection of holes into the silicon bar, and the resistance of R_{B1} increases to its original value, completing one cycle of operation.

The value of R_2 is very low and serves as the load across which the output pulse is taken. The waveforms in the circuit are shown in Fig. 7-6(B). The off-time of the pulse is determined by the time-constant of R_1C_1 while the on-time is a function of $(R_2 + R_{B1})C_1$. Therefore, the frequency of the oscillator may be varied by varying R_1.

The unijunction transistor oscillator offers several advantages over the multivibrator. The circuit is not dependent on supply voltage and hence shows greater frequency stability under supply voltage variations. Further, it is simpler in construction, requiring fewer components. However, its sharp pulse output

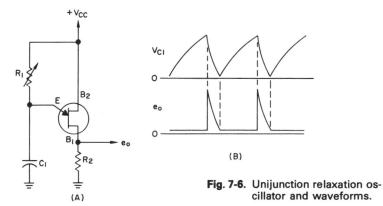

Fig. 7-6. Unijunction relaxation oscillator and waveforms.

must usually be converted to a more rectangular output for clock use. This is usually accomplished by feeding the pulse output of the unijunction oscillator into an R-S flip-flop.

Crystal-Controlled Oscillators

Greater frequency stability is obtained with crystal-controlled oscillator circuits. When the crystal is housed in a temperature-controlled housing, the greatest frequency stability is obtained.

Crystal-controlled oscillators are most often sine-wave oscillators, operating on feedback through the crystal. Since the crystal has a resonant frequency, determined primarily by its physical size, the feedback will occur only at this frequency. A typical circuit is shown in Fig. 7-7. The circuit employs two inverters (single-input gates), with the output of Q_1 fed through the crystal (xtal)

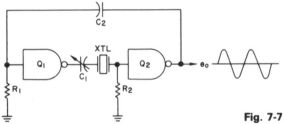

Fig. 7-7. Crystal-controlled oscillator.

to the input of Q_2. The output of Q_2 is fed to the input of Q_1 through C_2. Hence, the arrangement is a positive-feedback loop at the crystal's resonant frequency. C_1 provides some frequency adjustment. Resistors R_1 and R_2 change the operating point of each gate so that they operate as linear amplifiers. The output of this oscillator is a square wave since gates are employed. It is usually followed by a circuit, such as a Schmitt trigger, to develop a good square-wave or pulse output.

SIGNAL SHAPERS

The Ramp Generator

In many applications, it is necessary to convert the rectangular pulse signal into a ramp signal. The conversion from rectangular to ramp signals is accomplished by a capacitor charging circuit, as shown in Fig. 7-8. The rectangular pulse input turns the transistor on and off. When the input is at zero, the transistor is off and the capacitor charges toward $+ V_{cc}$ through R_L.

When the input is at its positive $(+)$ level, the transistor is turned on (saturation), the collector-to-emitter is at very low resistance, and the capacitor discharges quickly through this low resistance. The resultant waveform is shown in Fig. 7-8(B). The values of R_L and C are made large so that the capacitor charges over its initial and most linear portion of its charge cycle; this produces a ramp that is fairly linear.

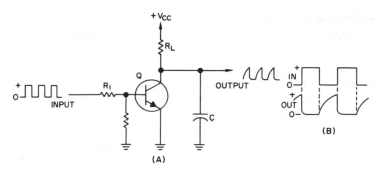

Fig. 7-8. Ramp generator circuit: (A) schematic, and (B) input/output waveforms.

A much more linear ramp is achieved if the capacitor is charged from a source which can present a constant charging current. Such a circuit is shown in Fig. 7-9(A).

Transistor Q_1 is a common-base amplifier whose collector current is approximately equal to its emitter current regardless of load. This current charges C_1 at a constant linear rate to $+V_{cc}$. This charging current is shown by the dashline in Fig. 7-9(A). Resistor R_1 sets Q_1's bias current to provide the desired amount of charging current passing from base to collector. A positive pulse at the base of Q_2 turns the transistor fully on, effectively short-circuiting C_1. C_1 quickly discharges to zero potential across its plates ($+V_{cc}$ on both plates) through Q_2 and also puts the output at $-V_{cc}$. When the input pulse ends, Q_2 turns off, and C_1 charges toward $+V_{cc}$. The output voltage is the algebraic sum of $+V_{cc}$ and the voltage across C_1. Since the two oppose one another, the output ramp will start at $+V_{cc}$ and decrease toward zero voltage as C_1 charges.

Transistor Q_1 can be considered as a *constant-current source*. The ramp waveforms are often used in digital voltage comparator circuits. More information on applications is presented in Chap. 10 in the discussion of digital voltmeters and multimeters.

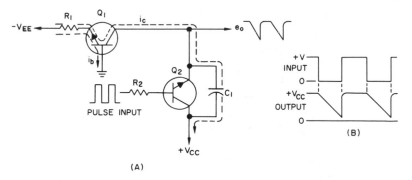

Fig. 7-9. Linear ramp generator circuit and waveforms.

The Staircase Generator

The circuit shown in Fig. 7-9 can also be used to develop a staircase waveform. This waveform is employed in many comparator operations. As shown in Fig. 7-10, if the constant-current source Q_1 is pulsed on at a regular rate, C_1 will charge to a level and will hold that level until the next pulse, when it will charge to the next level. The capacitor charges to five steps before transistor Q_2 is turned on, causing C_1 to discharge.

The staircase waveform can also be developed using a counter and digital-to-analog converter. Such an arrangement is shown in Fig. 10-17.

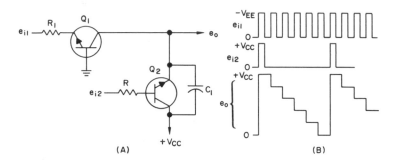

Fig. 7-10. A basic staircase signal generator.

The Schmitt Trigger

The Schmitt trigger circuit converts nonpulse-type waveforms to a pulse waveform suitable to drive digital logic circuits. For example, in a counter used to measure sine wave frequency, it is first necessary to convert the sine waveform to a pulse waveform before it is fed to the counter circuits.

The Schmitt trigger is basically an emitter-coupled, bistable multivibrator whose stable state is determined by the amplitude of the input signal. It, in effect, squares the input signal. The basic circuit is shown in Fig. 7-11.

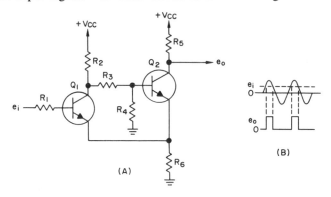

Fig. 7-11. Basic Schmitt trigger circuit and typical waveforms.

With no signal in, transistor Q_2 conducts to saturation, yielding an output of close to 0 level. The current flowing through R_6 develops a bias voltage to hold transistor Q_1 cut off. When a signal is fed in, Q_1 will turn on when the input signal level exceeds the bias voltage (E_{R6}). This in turn causes the voltage across R_4 to drop to a low level, and E_{R6} will now bias-off Q_2, causing the output to go to $+V_{cc}$. When the input signal drops below the E_{R6} level, Q_1 will turn off, causing Q_2 to turn back on and the output to drop to zero. If resistor R_6 is made variable, then the width of the output pulse can be varied.

The Schmitt trigger produces an excellent rectangular pulse and hence is often used as a conditioning circuit to clean up pulses from relay contacts, photoelectric sensors, etc., before these pulses are applied to other logic circuits. If these pulses were not cleaned up, they could cause errors in the logic system.

The One-Shot

The "one-shot" is a monostable multivibrator circuit which remains normally in one state. It can be switched to its other state for a period of time determined by the R-C time-constant of its feedback loop. The one-shot is very useful in providing pulses of a specific duration for gating or delay purposes. The basic one-shot circuit is shown in Fig. 7-12.

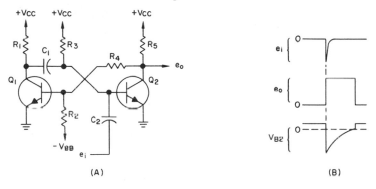

Fig. 7-12. Basic one-shot circuit and waveforms.

The circuit operates as follows: With no input, transistor Q_1 is biased-off ($-V_{bb}$ on base) and Q_2 is on. Capacitor C_1 charges to $+V_{cc}$. When a negative pulse is applied to the input, Q_2 is cut off, raising its collector voltage to $+V_{cc}$, forward-biasing Q_1. Q_1 turns on and conducts to saturation, effectively grounding the positive side of C_1. C_1's negative side now charges from $-V_{cc}$ toward $+V_{cc}$, through R_3, reverse-biasing Q_1 and holding it cut off. When C_1 has charged to zero voltage, Q_2 is no longer reverse-biased and comes out of cutoff into saturation, returning the circuit to its stable state. The time period during which Q_2 is held cut off is the time period of the output pulse.

The period of the pulse is a function of the size of C_1 and R_3. The period can be determined by $t = 0.69R_3C_1$.

For example, if $R_3 = 10$ kohms and $C_3 = 100$ pF, then:

$$t = 0.69 \times 10 \text{ k} \times 100 \text{ μμ} = 1 \text{ μsec}$$

If resistor R_3 is made adjustable, the width of the pulse can be varied.

Several manufacturers have available monostable multivibrator circuits in IC packages. The TTL 74121 family is an example.

Differentiator and Integrator Networks

R-C differentiator and integrator circuits are often used to shape a pulse waveform for desired circuit operation. The R-C differentiator circuit is used to change a rectangular pulse into a spike-shaped pulse. The R-C integrator circuit is used to change a rectangular pulse into a sawtooth pulse.

An R-C differentiator, as shown in Fig. 7-13, consists of a capacitor-resistor network functioning as a high-pass filter. The output is taken across the resistor. If the time-constant (RC) of the circuit is made one-tenth or less of the pulse duration time, the output will be differentiated into positive and negative spikes. The waveshape across the resistor occurs as the capacitor alternately charges and discharges rapidly. The differentiated waveform is used very often to trigger flip-flops, to ensure that the circuit switches only on either the leading or trailing edge of the pulse. This prevents any false triggering due to noise riding on the pulse.

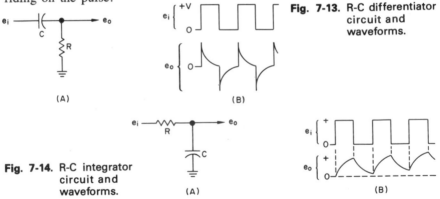

Fig. 7-13. R-C differentiator circuit and waveforms.

Fig. 7-14. R-C integrator circuit and waveforms.

The R-C integrator, shown in Fig. 7-14, consists of a capacitor and resistor in a low-pass filter arrangement. The output is taken across the capacitor. If the time-constant (RC) of the circuit is from one-tenth to ten times the duration of the pulse, the output will be integrated into the sawtooth waveform shown in Fig. 7-14(B).

Clippers and Clampers

Clipper circuits are employed when it is necessary to eliminate one portion of a wave while retaining another portion of the wave. Clamper circuits are employed to change the reference level of a waveform.

Figure 7-15 illustrates the operation of the basic clipper circuit. In Fig. 7-15(A), the diode in series with the input presents a low resistance on the positive alternation of the input and a high resistance on the negative alternation. Hence, current flows through R only on the positive alternation, clipping off the negative alternation. If the diode is reversed, the output will be the negative alternation.

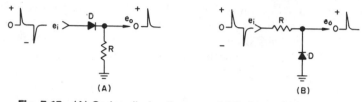

Fig. 7-15. (A) Series diode clipper, and (B) shunt diode clipper.

If the positions of the diode and resistor are interchanged, a shunt diode limiter circuit is obtained, as shown in Fig. 7-15(B). The diode is forward-biased on the negative alternation, shunting the signal to ground. On the positive alternation, the diode is reverse-biased, causing the signal to appear at the output.

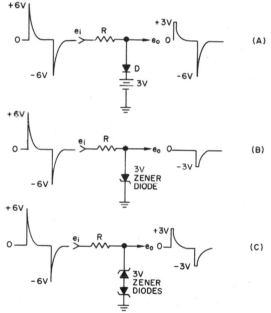

Fig. 7-16. Three examples of biased clipper circuits.

Clippers are sometimes biased using a voltage or a zener diode, so that only a portion of the wave is clipped. Several types of biased clipper circuits are shown in Fig. 7-16. In Fig. 7-16(A), the diode is reverse-biased until the input voltage exceeds +3 volts; hence, only this portion of the wave is clipped. In Fig.

7-16(B), the zener diode is forward-biased when the wave is positive, clipping off the positive alternation; when the wave exceeds -3 volts, the zener voltage is exceeded and the diode conducts, clipping the part of the signal which exceeds -3 volts. In Fig. 7-16 (C), the top zener diode is off until $+3$ volts is exceeded, while the bottom diode is off until -3 volts is exceeded. This circuit therefore limits the incoming signal to 6 volts peak-to-peak.

A basic clamper circuit is shown in Fig. 7-17. On the positive alternation, the diode is forward-biased, allowing capacitor C to charge to 5 volts. The output will be zero volts since the diode is essentially a short. On the negative alternation, the diode is reverse-biased, and the output becomes the algebraic

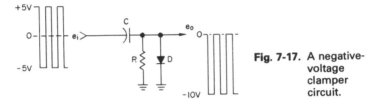

Fig. 7-17. A negative-voltage clamper circuit.

sum of the input signal level and the voltage across C. Since the charge on C $= -5$ volts and the input level is -5 volts, the output will be -10 volts. The resistor provides a discharge path for the capacitor so that if the input voltage level decreases in amplitude, the signal will still be clamped to zero. If the diode's polarity is reversed, the circuit will become a positive-voltage clamper.

CLOCK TIMING SYSTEMS

In a logic system, there are usually many logic circuits that must perform functions at specific times so that operations may be synchronized. The clock is therefore employed as a pulse source to synchronize these operations. Logic circuitry is then employed to create synchronized pulses for the various logic circuits. For example, let us say that we wish to depress a "start" switch to develop a start pulse that is synchronized to the clock. Figure 7-18 illustrates a method for accomplishing this.

As shown in Fig. 7-18, when the start switch is depressed momentarily, it creates a pulse which most usually will not be synchronized to the clock pulse. Further, due to contact bounce, it will contain noise pulses which could cause triggering errors. Therefore, the start pulse is fed to a one-shot circuit having a pulse period slightly less than two clock pulses. The one-shot also removes the noise pulses.

The output of one-shot A is fed to an AND gate. The other input to the AND gate is a differentiated pulse developed from the clock pulse. One-shot A therefore gates on the AND gate for a period sufficient to allow one positive differentiated pulse to pass through one-shot B. One-shot B will hence develop a start pulse that is now synchronized to the clock.

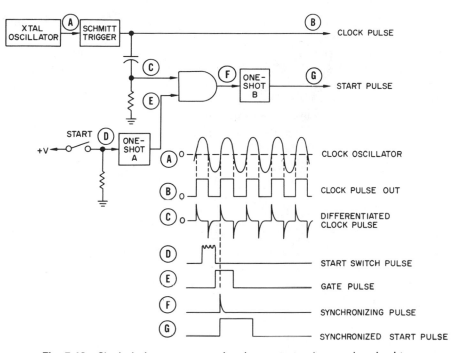

Fig. 7-18. Clock timing system to develop a start pulse synchronized to the clock pulse.

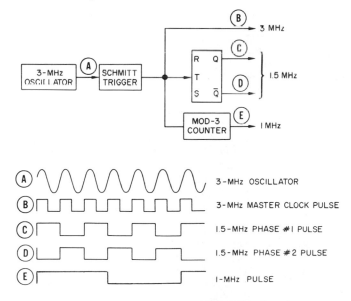

Fig. 7-19. Clock system for developing multiple clock pulse rates and phases.

Often it is necessary to develop clock pulses occurring at different rates than the master clock pulse. This is accomplished by dividing the master clock pulse rate. An example is shown in Fig. 7-19. The master oscillator develops a clock pulse at 3 MHz. This is then divided by an RST flip-flop to 1.5 MHz and by a Mod-3 counter to 1 MHz. Further, outputs are taken from the RST flip-flop's Q and $\overline{Q}$ outputs so that we have 1.5-MHz pulses occurring on alternate clock pulses. The 1.5-MHz pulses are 180° out-of-phase, and this output is often referred to as a *two-phase* output. Two-phase clock pulses are often employed in parallel counters to overcome race problems.

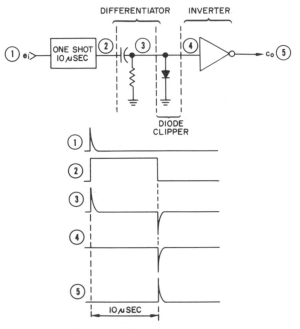

Fig. 7-20. Pulse delay circuit.

Frequently, it is desired to delay a particular timing pulse (e.g., the adder circuit shown in Fig. 6-7). The circuit shown in Fig. 7-20 is often used for this purpose. The one-shot circuit has a 20-μsec pulse period to provide a 20-μsec delay of the pulse. The one-shot's output is differentiated to provide positive and negative spikes. The leading-edge spike is clipped off by a diode clipper. The spike is then inverted to provide the delayed output. Delay times ranging from microseconds to minutes can be provided.

Review Questions

1. What is the advantage of a crystal oscillator compared to a multivibrator oscillator?

2. How can a sine wave oscillator's output be converted to a rectangular wave output?

3. What determines the frequency of a multivibrator?

4. If it is desired to decrease the pulse width in a Schmitt trigger circuit (Fig. 7-11), what must be changed?

5. Show by a block diagram, how a one-shot, differentiator, clipper, and inverter can be used to provide a delayed pulse. Show all waveforms versus time.

6. In a unijunction oscillator, what happens to frequency as R_1 is increased?

7. Draw the output waveshape of the circuit shown below.

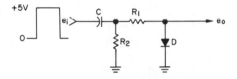

8. Draw the output waveshape of the circuit shown below.

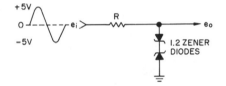

9. Draw the output waveshape of the circuit shown below.

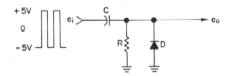

10. In the circuit shown in Fig. 7-19 explain how a 500-kHz clock pulse can be obtained.

Chapter 8

Memory Systems

At some point, in most digital computer systems, it is necessary to store data and/or instructions while operations are being performed in the operating registers. The date and/or instructions are stored in a memory. The magnetic core memory has been, and still is, the most widely used memory system. Its popularity is due to its very fast access time and ability to store information for long periods of time. Further, they have been around for a long time and their reliability is well proven. Although initially these units were quite expensive, improved manufacturing technology has reduced the cost substantially. However it still is not possible to manufacture them by automated techniques.

The semiconductor memory is a new entrant to the field. Although still more expensive than magnetic core for larger memories, it has taken over the smaller buffer memory and specialized memory areas. It holds promise, in the near future, of replacing core memory entirely. Semiconductors clearly have the advantage of high performance and small size.

Magnetic core and semiconductor memories offer the advantages of speed in accessing data into and out of memory. This feature is most desirable in the CPU (central processor unit) of a computer, where random access is required to data in the memory. However, substantial cost reductions can be effected when data and instructions are stored in bulk memories. The most popular types of bulk memories are magnetic tape, magnetic discs, magnetic drums, paper tape, and cards.

MAGNETIC CORE MEMORIES

The magnetic core memories find their widest application as *RAMs* (random-access memories) in CPUs of computers (central processor units, also known as "mainframes"). The typical density of core memories is around 2,000 bits/cu in. although core memories have been built with as much as 8,000 bits/cu in.

The heart of the core memory is the very small (typically 0.075 in. outside diameter) doughnut-shaped toroid made of ferrite and shown in Fig. 8-1.

It has high permeability so that it can be magnitized easily in either of two directions. When magnetized it can store a very high density of magnetic flux. When the direction of magnetic flux is changed, it produces a large output signal.

As shown in Fig. 8-1(A), if a wire passing through the core has a curent

Fig. 8-1. A magnetic memory core.

pulse passing through it, it will cause the core to be magnetized in a given direction. If the pulse of current is large enough, the core will reach saturation flux density. If the curent pulse is reversed, as shown in Fig. 8-1(B), the direction of the flux will change. Hence, we can say that a core can store a two-level binary bit by noting the direction in which it is magnetized. In other words, the core has two states; one is defined as the 1 state and the other as the 0 state. We will assume, in our discussions, that the $+I$ produces the 1 state and the $-I$ produces the 0 state. This current is called the *write* current since we write a 1 or 0 bit into the core by a current pulse passing through the wire.

The technique for *reading* the bit stored in the core requires a second wire, shown in Fig. 8-2. This wire is called the *sense* wire. We can detect the contents of the core by applying a $-I$ pulse to the write input to reset the core to the 0 state. If the core was in the 1 state, the switching of the core will induce a current in the sense wire. If the core was in the 0 state, no current is developed in the sense wire.

Magnetic cores are arranged in a matrix to form a memory system, as shown in Fig. 8-3. Each core has four wires passing through it. They are the X and Y select lines, and the common sense and inhibit lines. To write a 1 bit into a particular core, positive current pulses are fed into the appropriate X and Y select lines. Each current pulse is one-half the current necessary to switch the core to the 1 state; hence, a core is set to the 1 state only when its X and Y select lines have positive current pulses that are coincident (occur at the same time).

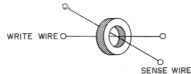

WRITE WIRE

SENSE WIRE

Fig. 8-2. The addition of the sense wire to read the contents of the core.

Each core is designated, or *addressed*, by selecting its respective select lines. Hence, we can designate a core at address X_1Y_1, X_2Y_1, X_2Y_2, and so forth. The designation X_1Y_1 becomes the address of a bit, since it specifies its location.

For example, we can put a 1 at address X_1Y_1 by putting coincident $+I$ pulses on the X_1Y_1 select lines. Since there are 16 cores in the matrix we can place 1's or 0's at any one of 16 addresses in the core matrix.

To read what is stored in the memory, $-I$ pulses are fed to the selected address to reset the core to 0. If a current pulse is produced on the sense line, then

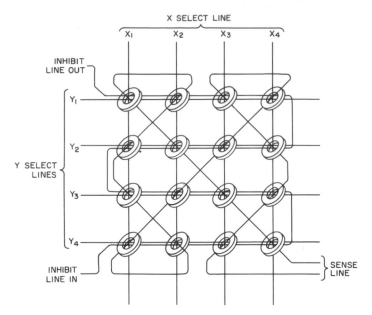

Fig. 8-3. A coincident-current memory plane (16 bits).

a 1 was present in the selected core. For example, to read the state of the X_1Y_1 core, coincident $-I$ pulses are fed into the X_1Y_1 select lines. If a pulse occurs on the sense wire during this time, a 1 was stored at X_1Y_1. If no pulse occurs, then a 0 was stored at X_1Y_1.

Since the sense wire passes through half the cores in one direction and through the other half in the other direction, the sense pulse may be either positive or negative. Hence, the sense pulse is amplified and rectified to produce an output pulse that is always of the same polarity.

A 0 is read into an address by pulsing the desired X and Y select lines while at the same time applying a half-current to the inhibit line (assuming that all cores were initially 0). The inhibit current cancels the additive effect of the X and Y select currents, inhibiting a 1 from being set into the selected core and leaving it set to 0.

The memory shown in Fig. 8-3 is one plane of a coincident-current core memory. A complete memory is created by stacking memory planes as shown in Fig. 8-4. All X select lines are connected in series and all Y select lines are connected in series. Each plane stores one bit in a word. For example, if we had four planes with 16 cores in each plane we would have a memory capable of storing 16 four-bit words. The total size of the memory can be said to be 64 bits, with 16 different addresses at which a 4-bit word can be stored.

If we increase the number of planes, we increase word size. If we increase the number of cores in each plane, we increase the number of words which can be stored. In the memory of Fig. 8-4, the most significant bit is stored in the bottom plane and the least significant bit in the top plane.

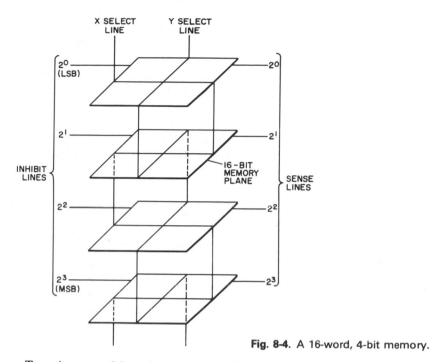

X SELECT LINE   Y SELECT LINE

2^0 (LSB)

2^0

2^1

2^1

INHIBIT LINES

16-BIT MEMORY PLANE

SENSE LINES

2^2

2^2

2^3 (MSB)

2^3

Fig. 8-4. A 16-word, 4-bit memory.

To write a word into the memory (all cores first set to 0), we select the desired X and Y lines (+ half-currents) while applying inhibit pulses to the planes where 0 is to be written. In the planes where a 1 is to be written, we do not apply inhibit pulses. In this manner, the word is written in memory in parallel form, all the bits at the same time.

To read a word stored at a particular address, we apply − half-currents to the designated X and Y select lines and detect whether pulses occur on the sense lines. This operation also causes the selected cores to be reset to 0, erasing the words from memory (clearing the selected address).

In actual practice, a write operation is preceded by a read operation to first clear the address. These operations are often referred to as the *memory cycle*. Further, since the word is erased from memory after each read operation, this system is called a *destructive readout* (DRO) system.

To increase the size of the memory, additional planes are added to increase word size, and more cores are added to each plane to increase addresses. For example, a memory to store 1,024 sixteen-bit words would consist of 16 memory planes with 1,024 cores in each plane.

Addressing the Memory

A word is located in memory by first loading its address, in binary number form, into the *memory address register* (MAR) as shown in Fig. 8-5. The MAR consists of four flip-flops, two for the X lines and two for the Y lines. This will permit the addressing of four X and four Y lines for a maximum of 16 locations in each plane.

The outputs of the MAR are fed to decoders which convert the binary coded address to X-Y select pulses. These decoders are essentially binary-to-decimal decoders similar to those shown in Chap. 3. The X-Y select pulses are fed through 2-input enabling gates to the X and Y select lines of the memory. Only when the read/write input is at a 1 level will the address be fed through to the memory.

An example of addressing the 16-word memory shown in Fig. 8-5 is as follows. Let us say that we wish to address the 13th word. The binary code for 13 is 1101. The binary coded address is serially fed into the MAR; the parallel outputs of the MAR are fed to the X and Y decoders and through the select gates to the memory. The binary address 1101 is decoded to select X_0 and Y_1 lines, thereby selecting the 13th core in each memory plane.

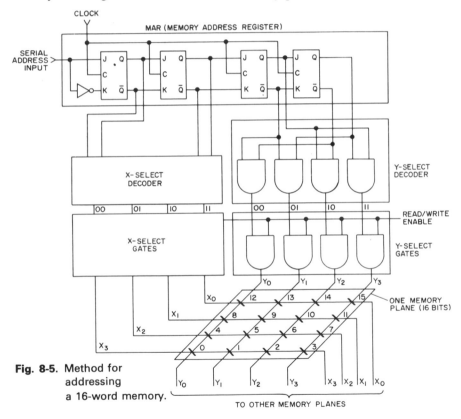

Fig. 8-5. Method for addressing a 16-word memory.

THE NDRO MEMORY

In the previous examples, the data stored in the memory was erased when a read operation was performed. The NDRO (nondestructive readout) system permits the reading of information stored in the memory while retaining the information in the memory. The circuitry for an NDRO system is shown in Fig. 8-6. The system shown also has provisions for writing new information into the

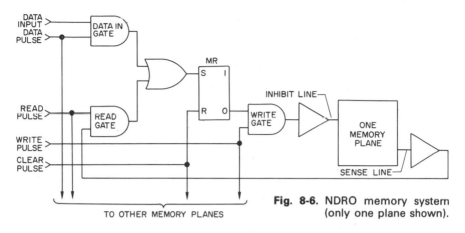

Fig. 8-6. NDRO memory system (only one plane shown).

TO OTHER MEMORY PLANES

memory. For clarity, the X-Y select circuitry is not shown.

The NDRO system utilizes a memory cycle during which a series of operations are performed in a prescribed sequence. The memory-cycle timing sequences (read and write) are shown in Fig. 8-7(A). A read memory cycle

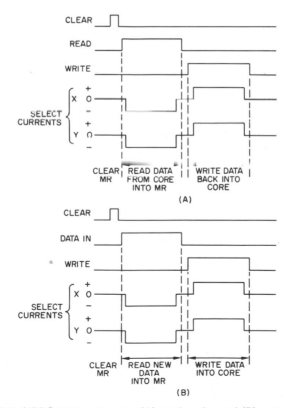

Fig. 8-7. NRDO timing signals: (A) read cycle, and (B) write cycle.

begins with a clear pulse, which resets the MR. The read pulse follows the clear pulse. During the read time, the X and Y reset pulses are applied, causing any of the selected cores which were in the 1 state to switch to the 0 state and generate a sense output. The sense amplifier amplifies the positive part of the sense output, and the read gate allows the pulse to be transferred through the OR gate into the MR where the data is stored.

The write pulse follows the read pulse. If a 1 is stored in the MR, the 0 output of the flip-flop will be 0 and the output of the write gate will be 0, disabling the inhibit input and allowing the X and Y set pulses to set the core back to a 1 state. If a 0 was in the MR, the 0 output of the MR flip-flop will be 1, enabling the inhibit line and preventing the core from being set to a 1.

A write cycle is shown in Fig. 8-7(B). The clear pulse resets the MR. The data-in pulse allows the new data to enter the MR, while the selected memory cores are set to 0. The write pulse and X and Y select pulses now set the data into memory in the same manner as previously described in the read cycle.

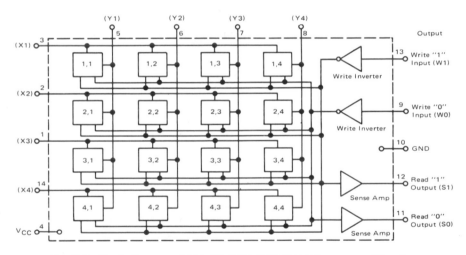

Fig. 8-8. Block diagram of MC-4004/MC-4005 IC-RAM memory. (*Courtesy Motorola*)

SEMICONDUCTOR IC MEMORIES

IC semiconductor RAMs have literally taken over in the applications of cache memory,* calculator memories and buffer memories. Further, the

---

*The cache memory was introduced by IBM in 1969. It is a submemory system. It is based on the theory that any word called up from the main memory will probably be used more than once before being returned to the main memory. Hence, words fresh from main memory and recently generated words are stored in the small, fast, temporary "cache memory." The last word entering the cache memory has the highest priority of retention. The oldest words are passed out of the cache memory as the newest enter. It has, in large memory systems, improved computation speed three to four times.

semiconductor ROM (read-only-memory) is widely used for code converters, character generators, look-up tables, microprograms, and keyboard encoders.

The storage element of the semiconductor memory is the bistable flip-flop (refer to Chap. 4). It may utilize either bipolar or MOS-type devices. A typical bipolar IC-RAM is shown in Figs. 8-8, 8-9, 8-10, and 8-11. The memory is a

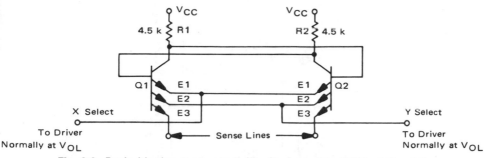

Fig. 8-9. Basic bipolar storage cell (flip-flop) used in RAM of Fig. 8-8. (*Courtesy Motorola*)

Motorola type MC-4004/4005 sixteen-bit RAM housed in a 14-pin plastic DIP case. The memory operates from a 5-volt supply and interfaces directly with TTL and DTL logic. The IC chip contains 16 flip-flop storage elements in a four-by-four matrix, as shown in Fig. 8-8.

A single bit of the matrix is selected by driving one of four X select lines and one of four Y select lines above the select threshold. The two sense amplifiers are shared by all 16 bits and provide a double-rail output from the selected bit. Two separate write amplifiers allow a 1 or 0 to be written into a selected bit.

Basic Circuit Operation

Each of the 16 storage flip-flops consist of two multiple-emitter transistors. The X select and Y select lines constitute two of the three emitter connections to each cell (Fig. 8-9) and are normally held at V_{OL} by the drivers. The third emitter (E_3) on each transistor is connected to one of two sense lines and is held by the associated sense amplifiers at approximately 1.5 volts.

To begin analysis of the operation of the basic storage flip-flop, assume that Q_1 is in the on (saturated) state. When this is the case, a 0 is stored in the flip-flop. Since the emitters E_1 and E_2 are biased at V_{OL} from the select drivers, the base of Q_1 will be at $V_{OL} + V_{BE}$. The collector of Q_1, and consequently the base of Q_2, will be biased at the V_{OL} of the driver plus the $V_{C\ E(sat)}$ of Q_1. Hence, Q_2 will be biased-off. As long as either the X or Y select line is below the select threshold voltage, the flip-flop will remain in its original state.

When the X and Y select lines of a flip-flop are simultaneously driven above the select threshold voltage, the E_3 emitter-base junction of Q_1 becomes forward-biased, and current flows out of E_3 into the sense line. In the read mode, this current is detected by the sense amplifier and the sense outputs are activated.

Digital Logic Circuits

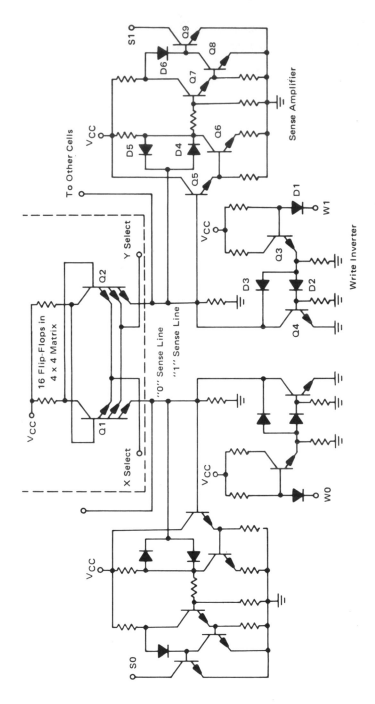

Fig. 8-10. Single storage cell with associated write and sense circuit used in RAM of Fig. 8-8. (*Courtesy Motorola*)

Write Operation

This is accomplished by lowering the voltage on the appropriate sense line while the bit is selected. A write amplifier is included on each chip (Fig. 8-10) to provide fan-out to the storage flip-flops.

Assume that Q_1 is on and Q_2 is off, and the select inputs of the bit in question are high. To write a 1 into the selected cell the W_1 line is driven high. Current is now routed through Q_3 and Q_4, and the collector voltage of Q_4 drops to approximately V_{BE}, biasing-on Q_2 to change the flip-flops state.

Read Operation

The sense amplifier biases the sense line at approximately 1.5 volts when no current from the memory cells is flowing in the associated sense line. The collector of Q_6 is biased at approximately $3 \times V_{BE}$; hence, Q_7 and Q_8 are biased-on, causing Q_9 to be off. When selection occurs, the current flowing in the sense line increases the base drive to both Q_5 and Q_6, causing the voltage at the collector of Q_6 to be lowered. Q_7 and Q_8 are turned off, causing Q_9 to turn on. Thus, when a 1 is stored in the selected bit, the S_1 output goes low, or, when a 0 is stored, the S_0 output goes low.

Figure 8-11 illustrates the application of the MC4004 in a RAM 16-word scratch pad memory system (DRO) of N bits. The system provides high-speed (read access time of 22 nsec), small size, and low cost.

MOS MEMORIES

The primary limitations of bipolar memories are size and power dissipation. As the size of the memory is increased, the drive and sense circuits become more complex and require greater power, increasing the size of the IC chip. The MOS memories offer the advantage of smaller cell size and lower power consumption, making higher cell density possible.

A typical MOS flip-flop memory cell is shown in Fig. 8-12. Q_2 and Q_4 serve as the loads for the flip-flop, consisting of Q_1 and Q_3. Q_5 and Q_6 are bidirectional switches (gates). The cell is read when the word line is high, enabling Q_5 and Q_6 and thereby transferring the data to the bit lines. To write into a cell, the word line is made high, enabling Q_5 and Q_6 and allowing the data on the bit lines to set the state of the flip-flop.

MOS devices are generally used also for decoding, and this is usually included on the same chip with the memory cells. Since the MOS memory is most often interfaced with TTL and DTL logic circuits, which require drive power, bipolar driving and sensing circuits are utilized.

A further reduction in power consumption and cell size and hence increased storage capacity on the IC chip is accomplished with charge-storage cell elements. The charge-storage element may be either an MOS transistor or bipolar diode. Present charge-storage memories utilize the 3-MOS device circuit such as is shown in Fig. 8-13.

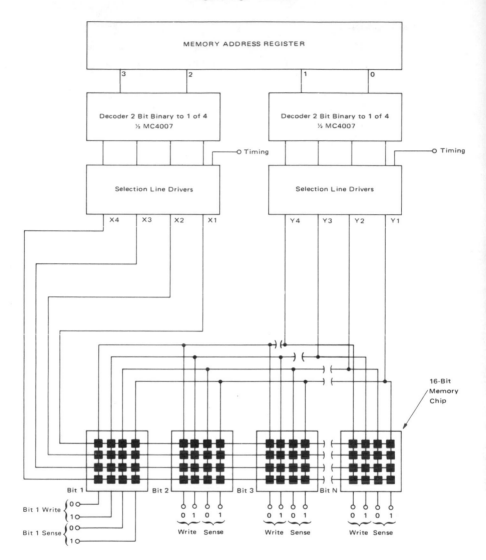

Fig. 8-11. A 16-word, N-bit scratch pad memory using IC-RAMs of Fig.
8-8. (*Courtesy Motorola*)

The MOS device has sufficient capacitance to store a charge on the gate
with respect to the channel. In a P-channel device, a negative charge on the gate
can represent a 0 and a positive charge on the gate can represent a 1. Q_1 is the
charge-storage element. Q_2 is the read gate and Q_3 is the write gate. Q_3 also
provides periodic refreshing of the cell. Refreshing is necessary since the charge
dissipates through the insulation resistance of the gate over a finite period. The
charge in Q_1 is rewritten, or reinforced, from an on-chip refresh amplifier.

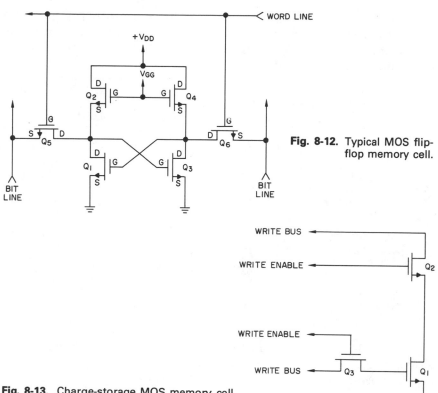

Fig. 8-12. Typical MOS flip-flop memory cell.

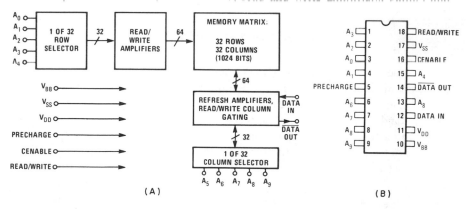

Fig. 8-13. Charge-storage MOS memory cell.

The type 1103 memory shown in Fig. 8-14 is a very popular 1,024-bit IC RAM utilizing a 3-MOS transistor-type memory cell. It also includes, within its 18-lead plastic DIP case, the decoders, read and write amplifiers, gating, and

Fig. 8-14. The type-1103 RAM-IC: (A) block diagram, and (B) IC pin connections. (*Courtesy Intel*)

refresh amplifiers. The block diagram of the 1103 is shown in Fig. 8-14. The unit features an access time of 300 nsec, an overall read/write time of 580 nsec, low power consumption (240 µW/bit selected, 6 µW/bit standby). The readout is nondestructive, and each cell is recharged once every 2 msec.

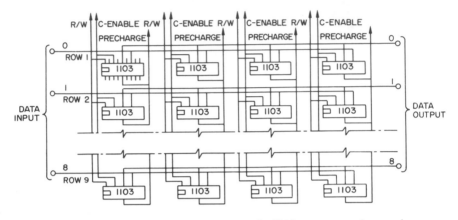

Fig. 8-15. Organization of a 4,096 word, 18-bit memory plane using 1103-RAMs.

The 1103-RAM is widely used as a buffer memory for CRT displays and as a mainframe memory in minicomputers. An example of a 4,096-word, 9-bit computer memory system is shown in Fig. 8-15 and 8-16. Figure 8-15 illustrates how the ICs are arranged in rows of four to provide 4,096 words and then nine rows (of four each) to provide nine bits (36 ICs total). Figure 8-16 shows the 36-IC memory and its associated control circuitry to form a complete NDRO memory system such as would be employed in a minicomputer. All of the circuitry is usually mounted on one printed circuit card, as shown in Fig. 8-17.

THE READ-ONLY MEMORY

A read-only-memory (ROM) contains information that cannot be altered. Its information is entered during the manufacturing process. The ROM can be either randomly addressed or sequentially addressed, depending upon the application. Random addressing is used in code translation applications, such as in a binary-to-memory address decoder and converting from one code to another. An example of the later is a Hollerith Code-to-ASCII code converter, shown in Fig. 8-18, which converts the 12-character Hollerith code to the 8-bit ASCII code. In the Hollerith code, two or more bits, in rows 1 through 7 (refer to the next chapter on punched cards) are never activated simultaneously, and hence these

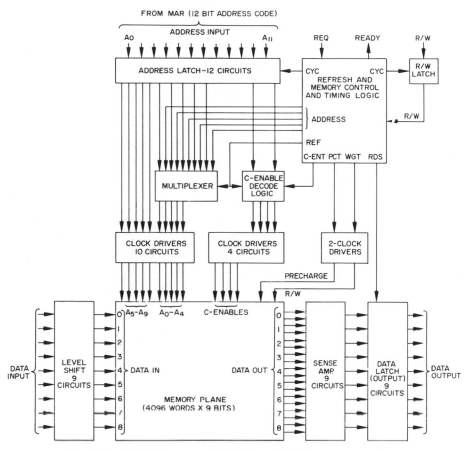

Fig. 8-16. Block diagram of a 4,096 word, 9-bit memory system.

bits are really a one-of-seven code which is compressed into a 3-bit binary code (A_2, A_1, and A_0) by a one-of-eight decoder (also a ROM). Therefore, only eight bits are fed into the ROM code converter.

A sequentially addressed ROM is often used in a computer system to store a microprogram or a subroutine such as the procedure for finding square roots. Often, when a computer is designed to perform only one objective, as in a process control system, the computer may not even have a RAM and utilizes an ROM instead. In these cases, the ROM is addressed sequentially by an address counter and the ROM generates the series of codes desired.

ROMs are also available that may be programmed and even reprogrammed after manufacture. These are referred to as PROMs (programmable read-only memory).

Fig. 8-17. An IC memory system printed circuit card. The system utilizes
the type-1103 RAM and provides 4,096 words × 9 bits. It
includes all driver, sense, and control circuitry. (*Courtesy Intel*)

BULK STORAGE MEMORIES

The memory in the CPU of a computer holds the data on which the
computer is currently working. Data is accessed one address at a time, from
memory. A bulk storage memory stores large amounts of data very economically, and large blocks of data are transferred when it is accessed. Bulk storage
memories are made using magnetic tape, disc and drum memories, paper cards,
and paper tape (the last two will be covered in the next chapter).

Magnetic Tape Storage

In the early days of computers, magnetic tape was the primary method of
bulk data storage. It is considerably more efficient than core memory in terms of
cost per bit and total storage space. Its access time is much slower than core
memory but faster than punched cards or paper tape. The data to be stored must
be organized into independent blocks with little cross referencing. When programs require skipping around from block to block, access time increases
substantially, as the time required to pass over irrelevant data blocks increases.
In this latter case, disc and drum storage are more efficient.

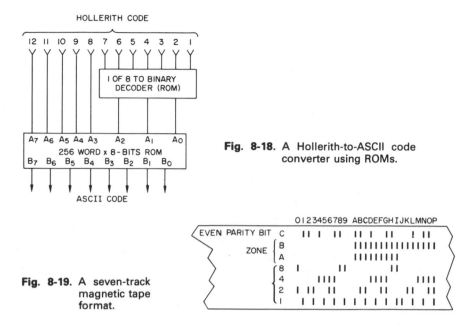

Fig. **8-18.** A Hollerith-to-ASCII code converter using ROMs.

Fig. **8-19.** A seven-track magnetic tape format.

Tape can also be erased and used over and over. Further, it is possible to store up to almost 400 million characters on one 2,400-foot reel of magnetic tape. The magnetic tape is most often used on 2,400-ft. reels and is a ½-in. strip of Mylar coated with magnetic metal oxide on one side.* The older systems employ seven-track (6 bits plus a parity-bit) formats, and the newer units use a nine-track format. The *A*merican *S*tandard *C*ode for *I*nformation *I*nterchange (ASCII, pronounced "as-key") uses six bits to form each character. The seven-track format is shown in Fig. 8-19. Each character consists of seven bits recorded on the tape by magnetizing the indicated spots on the tape. The lack of a spot (0) is either zero flux density or magnitized in the direction opposite to the spot (1).

The surface of the tape is magnetized by a special recording head, called a *read/write head,* which can write data or read data from the tape. A single-gap read/write head is shown in Fig. 8-20, and a two-gap head is shown in Fig. 8-21. The latter offers the ability to check the data on the tape as it is being recorded. As the tape passes across the gap, it is magnetized in one direction or the other. The magnetic polarity induced in the metal oxide denotes the 1's and 0's.

In the read operation, the magnetized area on the tape sets up a flux in the head as the tape passes over the gap, generating a small current in the winding. This pulse signal is amplified and fed through pulse-conditioning circuits. Data is usually written in blocks to reduce tape movement time, with inter-block gaps

---

*0.150-in. tape in cassettes has been recently introduced and is gaining acceptance due to its low cost.

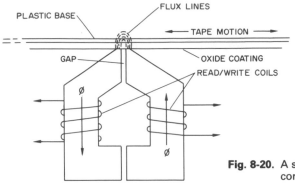

PLASTIC BASE

FLUX LINES

TAPE MOTION

GAP

OXIDE COATING

READ/WRITE COILS

$\emptyset$

$\emptyset$

Fig. 8-20. A single-gap magnetic tape recording head (read/write).

of about ¾ in. These gaps allow time for the tape to come up to speed and come
to a stop before and after reading or writing characters.

A typical magnetic tape unit is shown in Fig. 8-22. The read/write head is
located between the file and machine reels. The head assembly separates to
accept the tape for threading; when closed, the tape is in close contact with the
head for reading and writing. the tape is fed down through a vacuum column, up
past the head and down through another vacuum column to the other reel. The
loop of tape in each vacuum column is necessary to prevent the tape from
breaking during high-speed start and stop operations.

The independent action of the file and machine reels are made possible
through the use of vacuum-actuated switches located in the vacuum columns.
When the tape in the vacuum columns reach their maximum or minimum length,
either the file reel feeds tape, or the machine reel takes up tape, depending on
which vacuum column has reached its maximum or minimum length.

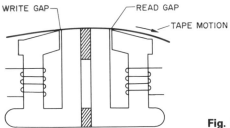

WRITE GAP

READ GAP

TAPE MOTION

Fig. 8-21. Two-gap read/write head.

Two methods of recording are used; RZ (return-to-zero) and NRZ
(nonreturn-to-zero). The RZ method is shown in Fig. 8-23. With no data in,
there is no current through the write head. A 1 is written on the tape by a positive
current pulse, and a 0 is written by a negative current pulse. After each pulse the
current returns to zero. The flux density on the tape is distorted due to tape
movement and frequency limitations of the system. When the magnetized areas
of the tape pass over the read head, they induce the currents shown on line 3 of

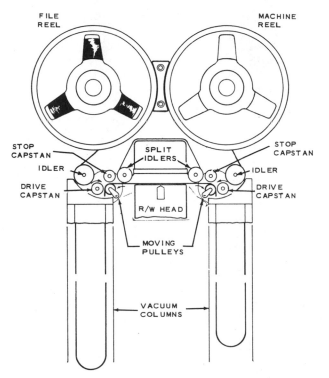

Fig. 8-22. Typical magnetic tape unit. (*Courtesy IBM*)

Fig. 8-23. Each recorded bit, whether a 1 or 0, will generate positive or negative pulses. To differentiate between the 1 and 0 pulses, a *strobe pulse* is used. The read head output, after being amplified, is fed, with the strobe pulse, to an AND gate. The timing of the strobe pulse is very critical to ensure proper sampling.

A further difficulty with the RZ method arises when recording over previous data. If the new data is not recorded precisely over the old data, the old data will remain on the tape, creating a conglomeration of data. The RZ method is employed effectively on magnetic drum memories where strobe pulses are recorded on one track of the drum, insuring proper timing.

One way of curing the problems of the RZ method is to use a biased-RZ method, as shown in Fig. 8-24. The write head current is normally at a negative current level and goes to a positive level for a 1. Now, when recording new data over old data, the tape is erased as the new data is recorded. This makes the strobe pulse timing less critical.

The NRZ method, shown in Fig. 8-24, records the data on the tape in the same manner as the biased-RZ method. The positive-going output of the read head is now sensed by feeding the amplified read head signal into an RS flip-flop, as shown in Fig. 8-25(B). The output of the flip-flop and the strobe pulse are then fed to an AND gate. The AND gate will produce a 1 only when a

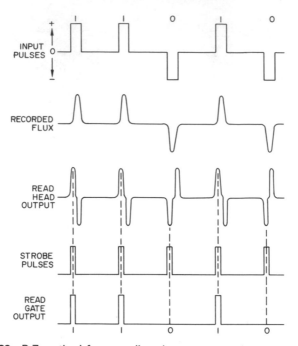

Fig. 8-23. R-Z method for recording data on magnetic tape or drum.

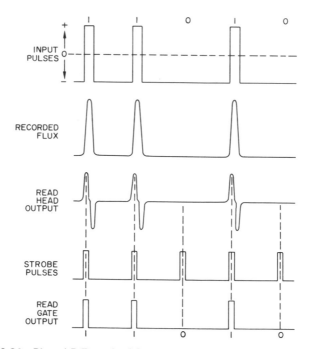

Fig. 8-24. Biased R-Z method for recording data on magnetic tape.

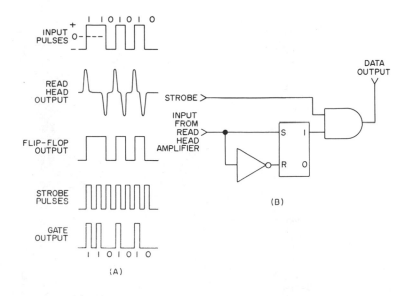

Fig. 8-25. NRZ method: (A) waveforms, and (B) logic circuit.

positive-going pulse is read from the tape. Notice now that the spacing between pulses is less critical, and hence greater bit density on the tape is possible than with the RZ method.

Disc and Magnetic Drum Memories

Magnetic drum and disc memories are bulk-storage memories that present a compromise between tape and core memories in terms of cost and access time. Typically, they can store about 300,000 bits/cu in. with access time of 2 msec. There are disc memory units currently available to store up to 3,200 megabits.

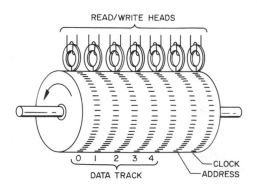

Fig. 8-26. Magnetic drum memory.

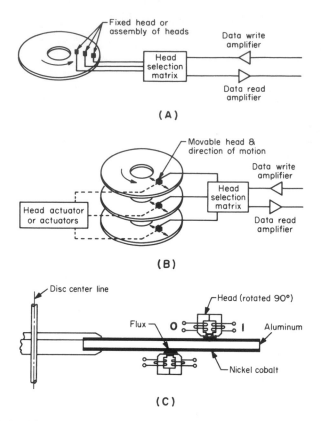

Fig. 8-27. Magnetic disc memory: (A) fixed head, (B) movable head, and (C) disc cross section. *(Courtesy Applied Magnetics Corp.)*

Magnetic Drum: The memory element, as shown in Fig. 8-26, is a metal cylinder coated with a magnetic material and rotating about the axis through its circular ends. A read/write head is used for each track. The RZ and NRZ methods are used for recording data. Since timing is critical, one track is usually used as a timing track and contains the strobe pulses.

Magnetic Disc File: The memory element here is one or a number of metal discs, coated on both sides with a magnetic material, as shown in Fig. 8-27. The disc rotates about the axis perpendicular to the centers of its flat side. Either a fixed head (one per track) or a moveable head is used. The head(s) are about 20–100 µin. above the disc, which can move at speeds up to 120 mph. The moveable head system is less expensive but requires a precision head-positioning assembly and has longer access time. Discs up to 40 in. in diameter are available. Both the RZ and NRZ recording methods are utilized.

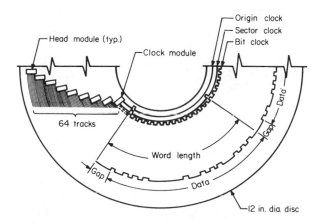

Fig. 8-28. Organization of a magnetic disc memory. (*Courtesy Applied Magnetics Corp.*)

The organization of a 12-in., 64-data track disc is shown in Fig. 8-28. The origin clock pulse occurs once every revolution. The sector clock pulse indicates the start of a word or block of data. The bit clock pulse indicates a bit and may occur from 16,000 to over 100,000 times per revolution. The clock pulses are permanently written on the disc.

Review Questions

1. Draw the construction of a coincident magnetic core memory having nine cores, and explain the operation.
2. Make a drawing showing the construction of a core memory capable of storing nine 6-bit words. Show all input and output lines.
3. How many bits can be stored in the memory of question 2?
4. Describe the difference between an ROM and an RAM.
5. Describe how information is organized on magnetic tape using the 7-track standard.
6. Describe two advantages and two disadvantages of magnetic tape storage.
7. Describe the two advantages and two disadvantages of drum storage.
8. Describe two advantages and two disadvantages of disc storage.
9. Describe typical applications of core, tape, drum, and disc memories.
10. Which recording method is more reliable, RZ or NRZ? Explain.

Chapter 9

Input/Output
Systems and Devices

How does one communicate with a computer or digital system? In other words, how can we get information into and out of the machine? This is the function of I/O devices (input/output) that take the language of the human communicator and convert it to the binary language of the digital system. The oldest and still most widely used input is the punched paper card (IBM card). But recent technological advances have led to other inputs, such as the teletypwriter and data terminal. Output from the digital system is most often a printer, with CRT displays, digital readouts, and X-Y plotters finding more and more application.

A typical computer system with input and output units is shown in Fig. 9-1. Input to the system is via punched cards. Output is either punched cards, printed forms, or storage on magnetic tape (refer Chap. 8).

PUNCHED CARDS

The punched card was invented by Hollerith in 1886, and, although modified and improved, it is basically the same as the original. A standard punched card using the Hollerith code is shown in Fig. 9-2. A binary code, with a hole representing a 1 and absence of a hole being a 0, is used to code standard alphanumeric characters. There are 12 rows and 80 columns. The card is capable of containing up to 80 characters. Each card contains one unit of information and is often referred to as a *unit record*, and the card-handling equipment is referred to as *unit-record equipment*.

Information is most usually entered on the card by a keypunch machine which has a typewriter-like keyboard. As the keypunch operator presses a key, the machine punches the holes for the code on the card. The information on the card is then entered into the computer by a card-reader machine at rates up to 100 to 1000 cards per minute.

The card reader, as shown in Fig. 9-3, reads one card at a time. As each card passes the read head, brushes or photocells sense a 1 or 0. There is a sensor for each column on the card. This information is fed to the input storage register of the computer. Very often a second read-head is used to verify the information taken from the card.

Very often the output from the computer is the punching of blank cards, and hence a reader/punch unit (Fig. 9-3) is frequently employed. Cards can be

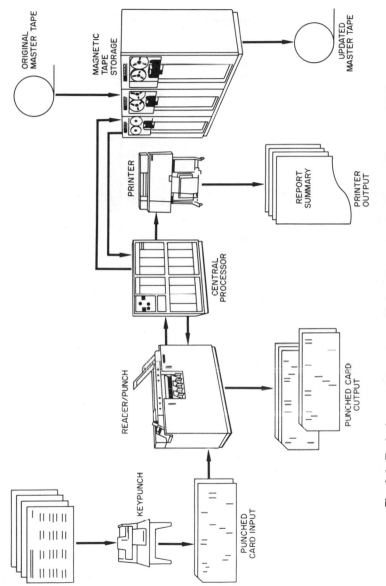

Fig. 9-1. Typical computer system with input/output units shown. *(Courtesy IBM)*

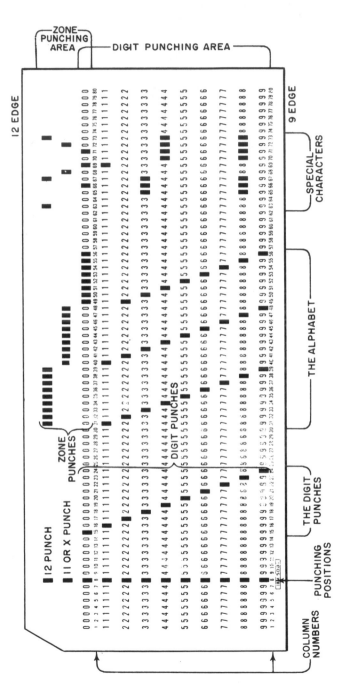

Fig. 9-2. Punched card using the Hollerith code. (*Courtesy IBM*)

punched at rates up to 250 cards a minute. The information will be in the binary "machine-code." The punched card is thus also useful as a memory to store information in bulk. It is easy to correct and change information by removing cards and inserting new cards.

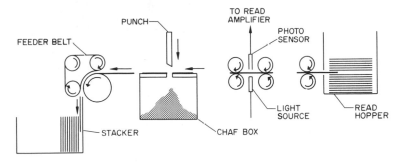

Fig. 9-3. Read/punch unit.

PAPER TAPE

Paper tape is a generic word and includes Mylar, Mylar-coated paper, and aluminum-coated Mylar tape. The information is put on the tape by punching holes in the tape. This medium has been in use for many years in communications work. It is widely used in numerical control and minicomputers where it is used to store programs. The paper tape program is generated as a byproduct of a keyboard operation or by the output of a computer. It offers the advantages of low cost, a faster input speed than punched cards, and high bit density. It can be corrected or edited with correction seals and splicing patches. Further, a user can tear off a tape and transmit it while preparing a second tape, which is not possible with magnetic tape. Also, a tape pertinent to a customer, for example, can be filed in a folder along with other related documents.

The tape punch is an electromechanical device, capable of punching up to 300 characters per second. Five-, six-, seven-, and eight-channel codes are used with the eight-level ASCII code (shown in Fig. 9-4), which is the American standard.

Tape readers use either spring-loaded pins, star wheels, or brushes to sense the holes. The star wheel, shown in Fig. 9-5, is widely used in low-speed readers. One wheel is used for each channel, and the presence of two successive holes or no holes in the channel allows the star wheel carrier arm and its corresponding contact to remain in position. When one of the points of the star wheel enters a hole, the respective switch is closed.

Photoelectric units generally use a single photovoltaic cell with nine separate channels for sensing. Photoelectric units can read up to 1,000 characters/sec. and have no contact with the tape to cause tape wear.

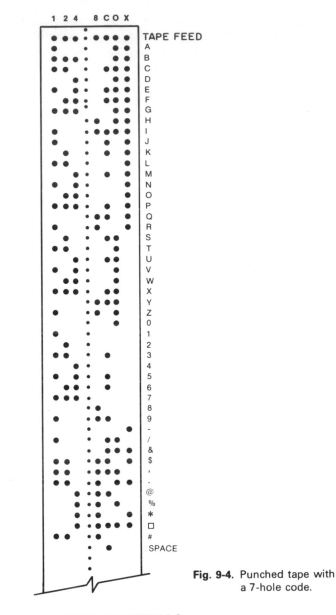

Fig. 9-4. Punched tape with a 7-hole code.

DATA TERMINALS

Data preparation and input accounts for more than 50% of data processing costs. The keypunch has dominated this area for 30 years, and today there are over 500,000 in use. During the past few years however, keyboard-to-tape and keyboard-to-disc units have gained in popularity due to the expensive processing

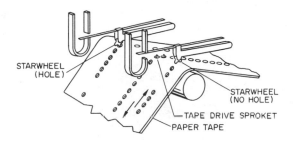

Fig. 9-5. Mechanical paper tape reader.

required by the keypunch. Further, the punching mechanism is so slow that a skilled operator can outpace the equipment. Also, cards are bulky and require large storage space, while one 2,400-ft reel of magnetic tape can store as much as can be stored on 32,000 punched cards.

Key-to-tape units record the data entered at the keyboard directly on magnetic tape, greatly simplifying data entry into the CPU of the computer. The teletype is a popular machine for generating the digital code. However, the introduction of the electronic keyboard has greatly reduced terminal cost and improved speed and reliability. It is apparent that the electronic keyboard will dominate the terminal area of the future.

Electronic keyboards utilize either mechanical contacts or solid-state switches, with the trend being toward the latter for increased reliability and lower cost. A typical mechanical switch is shown in Fig. 9-6(A). The key plunger has a magnet which actuates a reed-type switch as it moves from the up to the down position. A typical solid-state switch is shown in Fig. 9-6(B). When the key is depressed, the magnet is brought close to the integrated circuit, causing the increased magnetic flux to generate a voltage by the "Hall" effect. This turns the trigger circuit on.

The depressing of a key generates a binary code by means of an encoder. MOS-ROMs are used to generate the desired code. A typical circuit is shown in Fig. 9-7. The keyboard has 88 SPST switches and generates a nine-bit code. The ROM contains 792 bits, programmed to generate the code. When all the switches are open (all keys up), two ring counters (each of which have a single 1 always circulating) are clocked and sequentially address the ROM. When a key is depressed, there is a single path between one output of the 8-bit counter (X_0 through X_7) and one input of the 11-bit comparator (Y_0 through Y_{10}). After a number of clock cycles, the level on the selected path to the comparator will match the level on the corresponding comparator input from the 11-stage ring counter. At this time, the comparator generates a signal which, after being delayed, is presented at the strobe output. This strobe output indicates that valid data is present at the outputs of the encoder. Simultaneously, the comparator

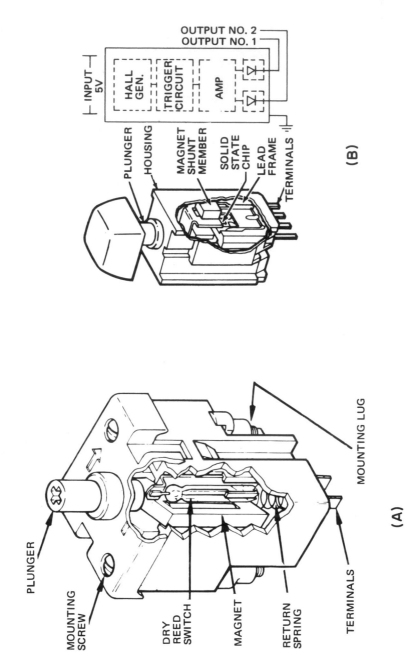

OUTPUT NO. 2
OUTPUT NO. 1

INPUT 5V — HALL GEN. — TRIGGER CIRCUIT — AMP

PLUNGER — HOUSING — MAGNET SHUNT MEMBER — SOLID STATE CHIP — LEAD FRAME — TERMINALS

(B)

PLUNGER

MOUNTING SCREW

DRY REED SWITCH

MAGNET

RETURN SPRING

TERMINALS

MOUNTING LUG

(A)

Fig. 9-6. Typical keyboard switches: (A) a mechanical reed switch, and (B) a solid-state switch. (*Courtesy Micro Switch*)

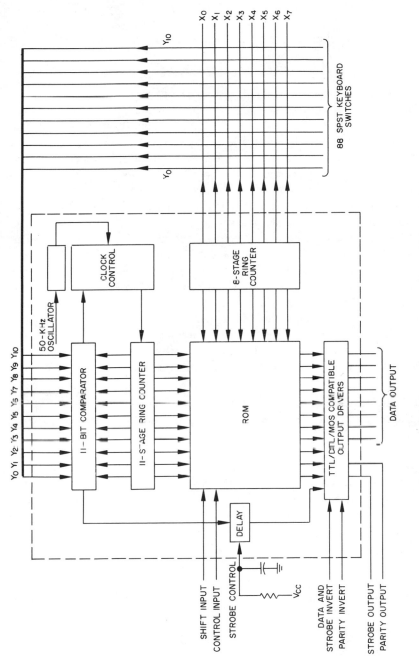

Fig. 9-7. An 88-key electronic keyboard with ROM encoder. *(Courtesy General Instruments Corp.)*

transmits a signal to the clock control which inhibits the clocks to the eight- and 11-bit counters. This stabilizes the data output until the key is released.

The most popular coding is the ASCII code shown in Fig. 9-8(A). The code used in IBM systems is the EBCDIC code shown in Fig. 9-8(B).

GRAPHIC DISPLAYS

Very often the data terminal will incorporate a graphic display in the form of a cathode-ray tube (CRT). This will allow the operator to see the information entered. Further, the CRT can also be used as an output for the computer. The more popular types of CRT displays utilize vertical or horizontal displays consisting of a matrix of rows and columns, with the excited segments combining to produce a complete character, as shown in Fig. 9-9. An MOS-ROM is used as a character generator to accept the character input code and present the character to the display, one column at a time. The example shown has a ROM which develops a 5 × 7 dot matrix.

A refresh memory is required both to buffer the incoming data and to refresh and/or update the display continually. The row memory is loaded from the page memory and recirculated for the number of times required to form a complete character row (or column). A new row is then loaded, and the process is repeated until the full raster area has been scanned. The sync and timing generator synchronizes the character generator with the CRT and contains counters for controlling the refresh memory and the ROM itself. The parallel-to-serial converter is used to control the on-off condition of the beam. The timing generator determines when the ROM outputs are to be sampled by the converter.

PRINTERS

The printer is the most widely used computer output device. It provides a permanent, continuous, readable record at a rate much higher than humans can accept. There are three broad categories: line printers, character printers, and strip printers. The line printer prints one entire line at a time at high speeds and is most usually used with computers. Character printers print one character at a time at low speeds and are most used in data terminals. Strip printers print one or more characters at a time with limited column width, at low speeds, and are most often used in data logging applications.

Line printers can produce up to 1,200 lines per minute. A typical line printer mechanism is shown in Fig. 9-10. It has a rated printing speed of 600 lines per minute and can print 48 different characters (26 alphabetic, 10 numerical, and 12 special characters). The characters are assembled in a chain which travels in a horizontal plane past the paper. Each character is printed as it passes ("on-the-fly") the selected position opposite a magnet-driven hammer that presses the hammer unit against the character chain.

ASCII CODE

Bits 4321 \ Bits 765 →	000	001	010	011	100	101	110	111
0000	NUL	DLE	SP†	0†	@	P	`	p
0001	SOH	DC1	!	1	A	Q	a	q
0010	STX	DC2	"	2	B	R	b	r
0011	ETX	DC3	#	3	C	S	c	s
0100	EOT	DC4	$	4	D	T	d	t
0101	ENQ	NAK	%	5	E	U	e	u
0110	ACK	SYN	&	6	F	V	f	v
0111	BEL	ETB	'	7	G	W	g	w
1000	BS	CAN	(	8	H	X	h	x
1001	HT	EM	)	9	I	Y	i	y
1010	LF	SUB	*	:	J	Z	j	z
1011	VT	ESC	+	;	K	[	k	{
1100	FF	FS	,	<	L	\	l	|
1101	CR	GS	-	=	M	]	m	}
1110	SO	RS	.	>	N	^	n	~
1111	SI	US	/	?	O	_†	o	DEL

† SHIFTED

(A)

EBCDIC CODE

Bits 4567 \ Bits 0123 →	0000	0001	0010	0011	0100	0101	0110	0111	1000	1001	1010	1011	1100	1101	1110	1111
0000	NUL	DLE	DS		SP	&	-						{	}		0
0001	SOH	DC1	SOS				/		a	j	~		A	J		1
0010	STX	DC2	FS						b	k	s		B	K	S	2
0011	ETX	DC3							c	l	t		C	L	T	3
0100	PF	RES	BYP	PN					d	m	u		D	M	U	4
0101	HT	NL	LF	RS					e	n	v		E	N	V	5
0110	LC	BS	ETB	UC					f	o	w		F	O	W	6
0111	DEL	IL	ESC	EOT					g	p	x		G	P	X	7
1000		CAN							h	q	y		H	Q	Y	8
1001	RLF	EM							i	r	z		I	R	Z	9
1010	SMM	CC	SM		¢	!	|	:								
1011	VT	CU1	CU2	CU3	.	$	,	#								
1100	FF	IFS		DC4	<	*	%	@								
1101	CR	IGS	ENQ	NAK	(	)	_	'								
1110	SO	IRS	ACK		+	;	>	=								
1111	SI	IUS	BEL	SUB	|	¬	?	"								LVM

(B)

Fig. 9-8. Popular keyboard codes: (A) ASCII, and (B) EBCDIC.

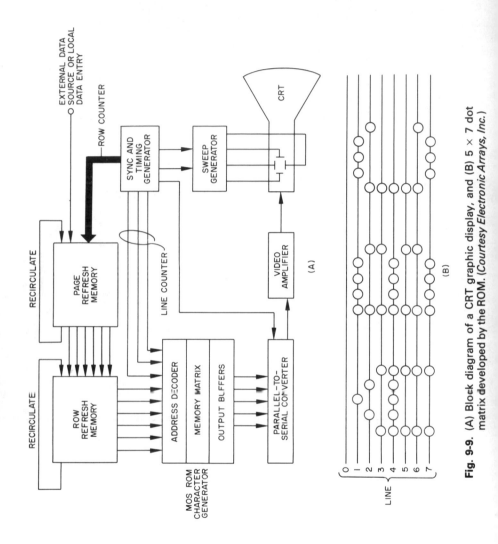

Fig. 9-9. (A) Block diagram of a CRT graphic display, and (B) 5 × 7 dot matrix developed by the ROM. (*Courtesy Electronic Arrays, Inc.*)

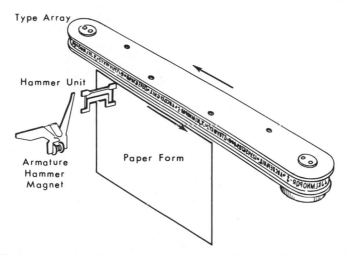

Fig. 9-10. A typical line-at-a-time printer mechanism. (*Courtesy IBM*)

MODEMS

The data modems (modulator/demodulator) are the interfacing devices which permit communications over voice-grade telephone lines between a data input device, such as a terminal, and a computer at another location. The modem modulates a tone carrier with the data stream. Three methods are used—FSK, ASK, or DPSK—with FSK being the most popular. Figure 9-11 illustrates the interconnection of devices.

Frequency-shift keying (FSK) involves frequency-modulating a tone carrier so that one frequency denotes a 1 and another denotes a 0. The data is in asynchronous form, as shown in Fig. 9-12. The data bits are grouped into

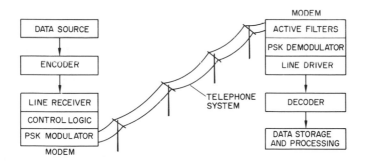

Fig. 9-11. Typical communication system using modems and telephone system.

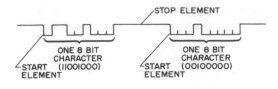

Fig. 9-12. Asynchronous data transmission.

fixed-length characters, and synchronizing start and stop elements are added to each character. A clock signal is not transmitted with the data. Rather, an independent clock is used at the receiving end. The FSK method is relatively immune to noise because the data is transmitted at a slow rate (up to 1800 bps). It is not possible to use it for high-speed transmission due to bandwidth limitations. The very widely used Bell system type-103 data set is an FSK modem. FSK units are generally used to connect teletypewriters on a time-sharing network.

Amplitude-shift keying (ASK) is used for high-speed transmission when vestigal sideband (VSB) is utilized. ASK provides very efficient utilization of the narrow bandwidth of the telephone system (nominally 4,000 Hz). Further, multilevel coding, as shown in Fig. 9-13, allows four possible levels, doubling the bit rate without increasing bandwidth requirements.

Phase-shift keying (PSK) is used primarily in high-speed data transmission. The most popular form is DPSK (differential phase-shift keying), shown in Fig. 9-13, where the phase angle of the carrier is shifted relative to the preceding angle rather than with respect to a reference signal. DPSK has a higher noise susceptibility than the preceding systems.

ASK and PSK permit transmission of data at speeds up to 9,600 bps over

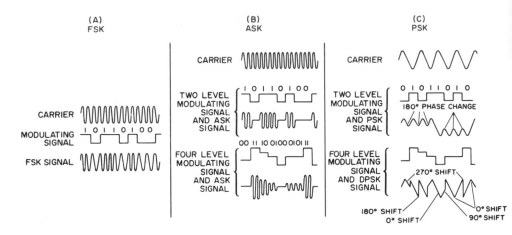

Fig. 9-13. Modern modulation techniques: (A) FSK, (B) ASK, (C) PSK.

standard telephone lines. They are therefore used in a computer-to-computer, tape-to-computer, and facsimile communications.

DIGITAL READOUTS

A wide variety of readout devices are available which permit direct visual output of the digital system. These readouts may be electromechanical, gas-discharge, light-emitting diode, liquid crystal, electroluminescent, incandescent, and on and on. The more widely used devices are the gas-discharge tube and light-emitting diodes (LED). (Due to space limitations we will confine our discussion to these two types.) The widest application of these devices is in calculators and digital measuring instruments such as voltmeters and frequency counters.

Fig. 9-14. Cold-cathode digital readout (Nixie R) with IC decoder and driver. (*Courtesy Burroughs Corp.*)

Gas-Discharge Display

More commonly referred to as Nixie®, the gas-discharge tube was introduced by the Burroughs Corporation in 1955. It is basically a cold-cathode tube (Fig. 9-14) with separate cathodes in the shape of characters. When a sufficient potential is applied (approximately 175 volts) between the selected cathode and plate, the gas surrounding the selected cathode is ionized and glows. Other types of gas-discharge displays are available having glow-bar segments to form in-plane characters and also cathode rods to illuminate a printed or cut-out character.

Figure 9-15 shows how a typical gas-discharge display tube is driven from TTL logic. IC-7490 is a decade counter with parallel BCD output and serial

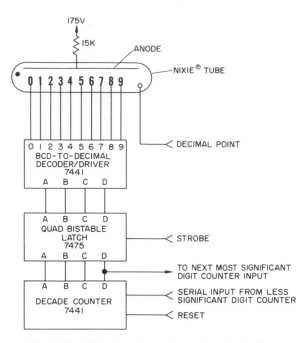

Fig. 9-15. Nixie R tube and associated circuitry.

count input. The overflow from the counter is used to operate the next more significant digit display. The BCD counter output is fed to a quad bistable latch (IC-7475) which acts as a memory storage. The count is stored in the IC until a strobe pulse allows the output of each latch to change. This prevents the display from being a blur of changing numbers. The BCD output of the latch is decoded by IC-7441 (BCD-to-decimal decoder/driver) to a one-of-ten decimal output. The IC also has high-voltage driver transistors to drive the cathodes of the tube.

Multidigit gas-discharge displays are also used, as exemplified by the Burrough's *Self-Scan®* display panel shown in Fig. 9-16. The panel has an array of 7-dot columns capable of displaying up to 256 full alphanumeric characters. The panel, generally includes the decoder, driver, memory, and scan circuitry. The panel has a glass plate with an array of holes sandwiched between top-wire anodes above the glass, cathode strips and bottom-wire anodes below it. The holes form cells for the neon in a 4 × 7 dot pattern. The cathode strips, one for each 7-dot column, are scanned, and the gas in a cell becomes visible.

Light-Emitting Diodes (LED)

LEDs are electrically very similar to semiconductor diodes. They use gallium arsenide phosphide (GaAsP) or gallium phosphide (GaP) which emits light at the junction with sufficient forward current. LEDs offer the advantages of being solid-state, operating from low voltages, small size, and long life.

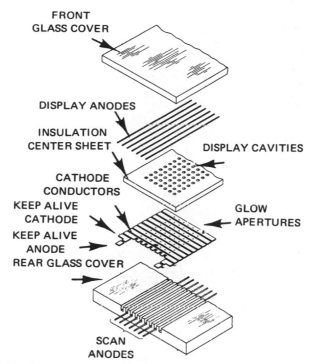

FRONT
GLASS COVER

DISPLAY ANODES

INSULATION
CENTER SHEET

DISPLAY CAVITIES

CATHODE
CONDUCTORS

KEEP ALIVE
CATHODE

KEEP ALIVE
ANODE

REAR GLASS COVER

GLOW
APERTURES

SCAN
ANODES

Fig. 9-16. Exploded view of Burroughs Self-Scan R multi-character display. (*Courtesy Burroughs Corp.*)

Typically, the LEDs are arranged in a 7-segment array. A typical LED numeric readout is shown in Fig. 9-17. It employs 28 diodes in a 7-segment array plus a "one" and decimal point. The decoder/driver is an LSI chip mounted on the same substrate with the diodes and then mounted in a hermetically sealed case.

A 5 × 7 LED matrix for alphanumeric readouts is shown in Fig. 9-18.

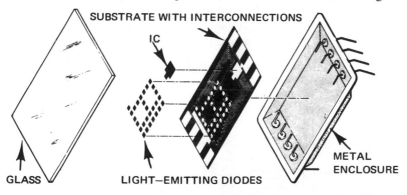

SUBSTRATE WITH INTERCONNECTIONS

IC

GLASS

LIGHT—EMITTING DIODES

METAL
ENCLOSURE

Fig. 9-17. Exploded view of LED digital readout. (*Courtesy Hewlett Packard*)

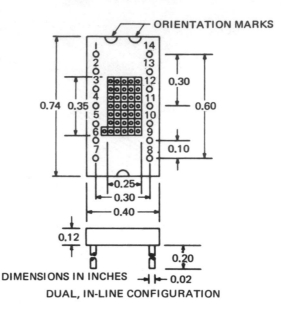
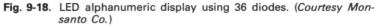

Fig. 9-18. LED alphanumeric display using 36 diodes. (*Courtesy Monsanto Co.*)

Review Questions

1. Show a section of paper tape encoded with the message ''Fox-25'' using the 8-hole code.
2. Show a sketch of a punched card with your name encoded on the card using the Hollerith code.
3. Explain the operation of a star-wheel-type paper tape reader.
4. If the digit ''8'' were transmitted serially using the EBCDIC code, draw the waveform it would create.
5. If the digit ''8'' were transmitted by a modem using four-level ASK technique with ASCII coding, show the waveform that would appear on the telephone line.

Chapter 10

D/A and A/D Conversion

Communications, once an exclusively analog technique, is rapidly changing to digital techniques. Telephones, interconnected by a pair of wires, transmitted sine-wave signals from handset to handset. Radio waves were modulated by sine-wave signals. Instrumentation such as voltmeters, temperature indicators, and motor controllers operated from continuously changing voltage levels. Today, this is all changing. Digitizing of analog information allows: (1) signals to be transmitted great distances without signal deterioration due to noise and distortion; (2) multiplexing of many signals on a common carrier with complete isolation of signals; (3) greater accuracy of signal measurement; and (4) processing of the signals through a digital computer.

Analog-to-digital (A/D) conversion is the process of converting an analog signal into a digital code. The A/D is sometimes called an encoder. The signal is said to be *digitized*. The process consists of assigning binary codes to the analog voltage levels.

Digital-to-analog (D/A) conversion is just the opposite. It is the process of converting the digital information into the analog signal.

For example, the telephone signal may be converted into a digitally coded signal by an A/D converter (ADC). The digital signal may be multiplexed with many other digital signals and all transmitted over one pair of telephone lines. This reduces the cost of installing wire pairs for each customer and the maintenance of the lines. Further, the digitized signal can be recreated by passing the signal through a logic gate, eliminating any noise or distortion introduced by the transmission system. At the receiving point, the signal is converted, by a D/A converter (DAC), back to an analog signal to drive the receiver.

D/A CONVERSION

The heart of the D/A converter is a resistive network which adds the pulses in the digital signal so that the sum of the pulses is equal to an analog level. Two different types of resistive networks are used: (1) the weighted divider, and (2) the binary ladder.

The Weighted Resistive Divider

In this system, binary weights are assigned. For example, consider the 3-bit binary numbers shown in Table 10-1. It is possible to have eight different binary coded numbers using the 3-bits. If we assign voltage levels to each code, we would begin by setting 000 equal to 0 volts and 111 equal to V_{max}.

Table 10-1. 3-Bit Binary Numbers Have 8 Distinct Codes.

Binary Bits →	MSB		LSB
	2^2	2^1	2^0
	0	0	0
	0	0	1
	0	1	0
	0	1	1
	1	0	0
	1	0	1
	1	1	0
	1	1	1

Since we have seven distinct coded numbers, in addition to 000, we can divide the binary numbers into seven voltage levels. Therefore, we can assign the 2^0-bit a weight of $1/7$, the 2^1-bit a weight of $2/7$, and 2^2-bit a weight of $4/7$. Table 10-2 shows these binary-weighted voltages. A binary number of 011 will sum to $3/7$ E_{max}.

Table 10-2. Weighted Voltages Produced by a 3-Bit Binary Code.

Binary Weights →	MSB		LSB	Sum of
Binary Bits →	$4/7$	$2/7$	$1/7$	Weighted
	2^2	2^1	2^0	Voltages
	0	0	0	0
	0	0	1	$1/7$
	0	1	0	$2/7$
	0	1	1	$3/7$
	1	0	0	$4/7$
	1	0	1	$5/7$
	1	1	0	$6/7$
	1	1	1	$7/7$

binary bits ⟶ $\quad 2^2 \quad 2^1 \quad 2^0$

binary number ⟶ $\quad 0 \quad\quad 1 \quad\quad 1$

binary weights ⟶ $\quad 0 + 2/7 + 1/7 = 3/7\ E_{max}$

A binary number of 101 will sum to $5/7$ to E_{max}.

$$2^2 \quad\quad 2^1 \quad\quad 2^0$$

$$1 \quad\quad 0 \quad\quad 1$$

$$4/7 + 0 + 1/7 = 5/7\ E_{max}$$

A binary number of 111 will sum to E_{max}.

$$2^2 \quad 2^1 \quad 2^0$$

$$1 \quad \quad 1 \quad \quad 1$$

$$^4/_7 + {}^2/_7 + {}^1/_7 = {}^7/_7 = 1 \; E_{max}$$

The sum of the binary weights must equal one. Hence, in a 3-bit binary code, $^4/_7 + {}^2/_7 + {}^1/_7 = {}^7/_7 = 1$.

If a 4-bit code is used in the D/A converter, there will be 16 codes and the weights will be $^1/_{15}$, $^2/_{15}$, $^4/_{15}$, $^8/_{15}$, as shown in Table 10-3. If $E_{max} = 15$ volts then $0001 = 1$ volt, $0010 = 2$ volts, $0011 = 3$ volts, and so on, up to $1111 = 15$ volts.

Table 10-3. Output Voltages Produced by a 4-Bit Binary Code.

Binary Weights →	MSB $^8/_{15}$	$^4/_{15}$	$^2/_{15}$	LSB $^1/_{15}$	Sum of Weighted	Analog Output
Binary Bits →	2^3	2^2	2^1	2^0	Voltages	Voltage (E_{max} = 15 volts)
	0	0	0	0	0	0 V
	0	0	0	1	$^1/_{15}$	1 V
	0	0	1	0	$^2/_{15}$	2 V
	0	0	1	1	$^3/_{15}$	3 V
	0	1	0	0	$^4/_{15}$	4 V
	0	1	0	1	$^5/_{15}$	5 V
	0	1	1	0	$^6/_{15}$	6 V
	0	1	1	1	$^7/_{15}$	7 V
	1	0	0	0	$^8/_{15}$	8 V
	1	0	0	1	$^9/_{15}$	9 V
	1	0	1	0	$^{10}/_{15}$	10 V
	1	0	1	1	$^{11}/_{18}$	11 V
	1	1	0	0	$^{12}/_{15}$	12 V
	1	1	0	1	$^{13}/_{15}$	13 V
	1	1	1	0	$^{14}/_{15}$	14 V
	1	1	1	1	$^{15}/_{15}$	15 V

A binary-scaling resistor network for summing a 3-bit binary number is shown in Fig. 10-1. The outputs of a storage register, or counter, are connected together through the network. R_L is the common load for all the inputs and is a very high resistance which does not load the input. If the 1 level of each flip-flop is 7 volts and $R = 1,000$ ohms ($R/2 = 500$ ohms and $R/4 = 250$ ohms), the operation of the circuit is as follows:

If the digital code is 001, as shown in Fig. 10-2(A), then the output of FF-2^2 and FF-2^1 will be zero (ground potential) and FF-2^0 will be 7 volts. Since R_L is very large compared to the other resistors in the circuit, virtually no current flows through it (and it can be ignored). The equivalent circuit then is shown in Fig. 10-2(B). The resistance of the 250- and 500-ohm resistors in parallel is 166 ohms, as shown in Fig. 10-2(C). The 7-volt input will therefore be divided so that 1 volt drops across the 166-ohm resistance and 6 volts drops across the

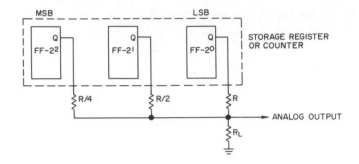

Fig. 10-1. A binary-scaled resistor network for summing a binary number.

1,000-ohm resistance. Since the output is taken with respect to ground, the output voltage will be 1 volt.

If the binary code is 101, the summing network will operate as shown in Fig. 10-3. The 2^1 flip-flop is at 0 (ground), while flip-flop-2^0 and flip-flop-2^2 are at 1 (7V) as shown in Fig. 10-3(A). The equivalent circuits are shown in Fig. 10-3(B) and (C). The 7 volts will be divided so that the output will be 5 volts. The network hence divides the 7-volt input in accordance with the division shown in Table 10-2 to provide 8 distinct levels from zero to 7 volts.

If a 4-bit resistor network is used, the circuit will be as shown in Fig. 10-4. The LSB is now eight times the value of resistor R. The voltage (1 level) will now be divided to provide 16 distinct voltage levels, as indicated in Table 10-3. As the number of bits is increased, the MSB resistor becomes less and less. This is a

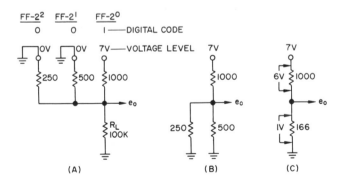

Fig. 10-2. Operation of the binary-scaled resistor network, code = 001.

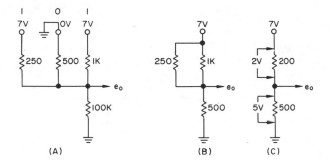

Fig. 10-3. Operation of binary-scaled resistor network, code = 101.

serious disadvantage in DAC's having a large number of bits, since the smaller the resistor the larger the load current. The MSB flip-flop must be capable of providing this large current.

A second disadvantage of this network is that resistor tolerance is very critical. In order for the DAC to be able to detect the difference of an LSB change at the input, the resistors must have a very tight tolerance. For example, in Fig. 10-4, if the MSB resistor is off by as little as 1 per cent, that input alone would contribute almost as much voltage change as the LSB. Therefore, resistor accuracy and temperature coefficients are very important to achieve accuracy.

The binary-weighted resistor network is popular in DACs up to 8 bits. However, for large systems the size and tolerance of the resistors become prohibitive. With the new state of the art in solid-state resistors, another resistor network technique has become popular. It is called the R-2R ladder network.

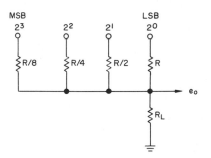

Fig. 10-4. A 4-bit binary-scaled resistor summing network.

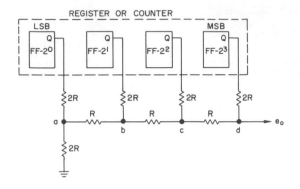

Fig. 10-5. A 4-bit, R-2R binary ladder resistor network.

R-2R Ladder Network

It is desirable to have a resistive network whose output impedance is constant regardless of the number of stages, maintains the binary weighting of the input signals, and uses equal-value resistors rather than many different values. Further, using resistors of the same value results in a cost reduction by purchasing them in high volume.

A four-bit R-2R ladder network provides these advantages and is shown in Fig. 10-5. The ladder network has the appearance of a transmission line made up of discrete resistors so that the 2R resistor from point A to ground terminates the ladder network in its characteristic impedance, thus eliminating impedance mismatching. Each flip-flop output therefore sees the same resistive load, namely 2R. This can be observed if each digital input is put at 0 (ground). Now, the resistance to ground at point A is 2R. The resistance at point B consists of two 2R resistors in parallel (equal to R) and one R resistor in series, as shown in Figure 10-6, making the equivalent resistance 2R at point B. In the same manner, the resistance at points C and D are also 2R.

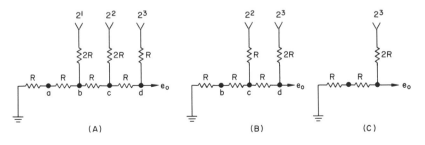

Fig. 10-6. The resistance along the R-2R ladder is seen to be constant at 2R.

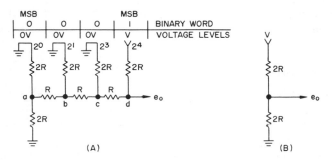

Fig. 10-7. The R-2R ladder network with a binary word input of 0001.

If a binary word of 0001 were entered into the network, the conditions would be as shown in Fig. 10-7. If the input voltage were designated V, the output voltage (E_{out}) would therefore be ½ *V*. If only the 2^2 input were 1, the voltage point C would be ½ *V* and divided so that $E_{out} = $ ¼ *V*. Continuing this analysis, we can see that when the 2^1 input is 1, $E_{out} = $ $^1/_8$ *V*, and when the 2^0 input is 1, $E_{out} = $ $^1/_{16}$ *V*. Hence, a properly weighted analog output is achieved for the binary input.

Table 10-4. Binary Weights for an 8-Bit Binary-Coded Word.

	Bit Position	Binary Weights
MSB	2^7	$^1/_2$
	2^6	$^1/_4$
	2^5	$^1/_8$
	2^4	$^1/_{16}$
	2^3	$^1/_{32}$
	2^2	$^1/_{64}$
	2^1	$^1/_{128}$
LSB	2^0	$^1/_{256}$

An 8-bit DAC would have the weights shown in Table 10-4. If *V* = 10 volts and the binary word were 00101011, the analog output could be calculated as follows:

$$
\begin{aligned}
\text{MSB} \quad 2^7 &= 0 \\
2^6 &= 0 \\
2^5 &= 1 = {}^1/_8 \times 10 \text{ V} && = 1.250 \text{ V} \\
2^4 &= 0 \\
2^3 &= 1 = {}^1/_{32} \times 10 \text{ V} && = 0.321 \text{ V} \\
2^2 &= 0 \\
2^1 &= 1 = {}^1/_{128} \times 10 \text{ V} && = 0.078 \text{ V} \\
\text{LSB} \quad 2^0 &= 1 = {}^1/_{256} \times 10 \text{ V} && = \underline{0.039} \text{ V} \\
& && E_{out} = 1.688 \text{ V}
\end{aligned}
$$

If all the binary inputs were high, the output would approach 10 volts, but not quite reach it. The more bits used, the closer the maximum output voltage will come to V.

The D/A Converter

A complete 4-bit DAC is shown in Fig. 10-8. The binary word is entered into the storage register in parallel form. The storage register consists of R-S flip-flops which are set to 0 or 1 by the read-in pulse. For example, if the word is 0101 at the input, the 2^1 and 2^3 flip-flops will be set to 0 and the 2^0 and 2^2 flip-flops will be set to 1 when the read-in pulse occurs. The output of the register is fed through level amplifiers which ensure that a 0 output is zero voltage, while a 1 output is at the precise V voltage. The outputs of the level amplifiers are summed in a binary-weighted resistor ladder network coupled to an operational amplifier (op-amp). An op-amp has very high gain and high input resistance. The characteristics of the op-amp are a function of the feedback resistor R_f. If R_f is equal to the source resistance (level amplifiers and ladder network), the resulting output voltage will be same as the input voltage. The op-amp has very low output resistance suitable for driving additional circuitry. Resistor R_c cancels the effect of the input bias current.

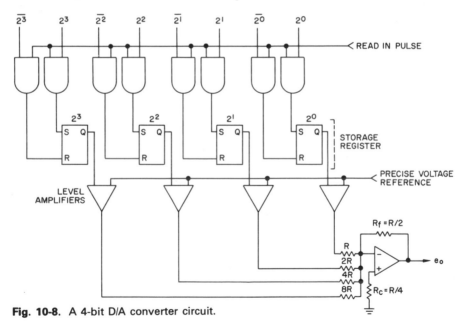

Fig. 10-8. A 4-bit D/A converter circuit.

A/D CONVERTERS

There are many analog-to-digital conversion techniques. The systems discussed here are most frequently used systems.

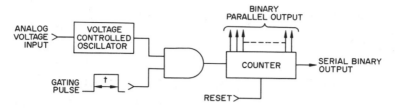

Fig. 10-9. Voltage-to-frequency A/D converter.

Analog-to-Frequency ADC

This method is one of the most popular due to its simplicity. The block diagram of the technique is shown in Fig. 10-9. The counter is initially reset to 0. The analog input is applied to a VCO (voltage-controlled oscillator) whose output is fed to an AND gate. The gate is enabled by a gating pulse from a fixed-frequency clock. When the gate is enabled, the counter counts the VCO's frequency for the gating interval.

For example, if a 1-sec gating interval is used and the VCO's frequency is 1 Hz with a 1-volt input, the counter will count 1 pulse during the gating period. As the analog voltage input is increased, the VCO's frequency increases. For example, at 20 volts input, the VCO's frequency would be 20 Hz and the counter would count 20 pulses, producing a binary output of 10100.

This type of ADC has a slow conversion time because of the fixed gating time, and the output code then represents the averge of the input voltage during the gating period. Accuracy is a function of the linearity of the VCO over the entire analog input range and the precision of the gating-pulse interval.

The Simultaneous ADC Method

This method is popular where ultra-high-speed ADC is required (as when digitally coded signals are multiplexed). The method, however, requires many comparators, which until the introduction of IC's were economically prohibitive. A 3-bit simultaneous ADC is shown in Fig. 10-10.

A parallel bank of seven voltage comparators is used to compare the analog input to seven different voltage levels. The comparator is a very-high-gain differential amplifier and therefore will go from cutoff to saturation (or vice versa) if a very small difference in voltage exists between its negative and positive inputs. For example, in Fig. 10-10, if comparator C_1 has a reference voltage of $^1/_8$ V and its input is less than $^1/_8$ V, its output will be 0. If C_1's input exceeds $^1/_8$ V, C_1's output will go to 1. Therefore, as the analog input is increased toward the voltage, each comparator, in turn, will produce a 1 output.

To understand the operation of the entire circuit, let us say that V is 10 volts and that the analog input is 5 volts. Then the outputs of comparators C_1 through

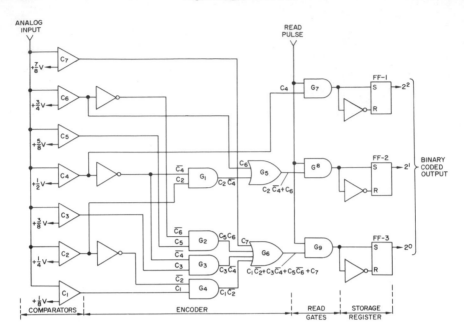

Fig. 10-10. A 3-bit simultaneous A/D converter.

C_4 will be 1 and the rest 0. The comparator outputs will be encoded by the inverters and gates G_1 through G_6, and fed to the read gates G_7 through G_9. In this case, the inputs to the read gates will be 100.

The code will be read into the storage register every time a read pulse is fed to gates G_7 through G_9. The storage register will have a parallel output.

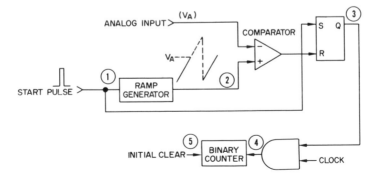

Fig. 10-11. Ramp-type A/D converter.

The Ramp-Type ADC

This method is simple in circuitry and provides medium accuracy, but is slow. It is popular because of its low cost. The block diagram is shown in Fig. 10-11.

The counter is first cleared. The start pulse actuates a linear ramp (sawtooth) signal, and sets the R-S flip-flop high, enabling the AND gate and allowing the clock pulses to be counted by the binary counter. When the ramp voltage reaches the level of the input analog voltage, the comparator switches states, resetting R-S flip-flop and causing the clock to be inhibited. The output digital count is now proportional to the analog voltage (V_A). The timing diagram for the circuit is shown in Fig. 10-12.

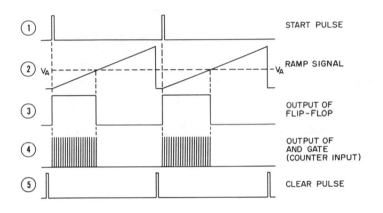

Fig. 10-12. Timing diagram for ramp-type A/D converter.

The Binary Ramp ADC

This method is the same as the previous one, with the exception that the ramp generator is replaced by a binary counter and a D/A converter (DAC). The block diagram is shown in Fig. 10-13.

The counter is initially set to zero count and allowed to count until the DAC output equals or exceeds the analog input, at which time the comparator changes state. At this point, the gate is disabled, thus inhibiting the clock pulses from reaching the counter. The binary output is taken from the counter.

This method is very good for high-resolution systems since as the number of bits is increased, very little additional circuitry is needed. However, as more resoltuion is required, the conversion time will increase rapidly.

The Continuous ADC

The primary disadvantage of the preceding method is that the counter must

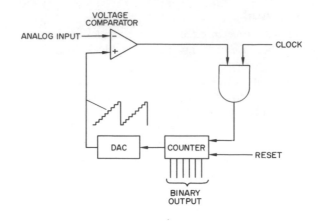

Fig. 10-13. Basic binary ramp A/D converter.

be reset to zero before each count, causing a long conversion time. The continuous ADC, shown in Fig. 10-14, overcomes this limitation by using an up/down counter so that the counter need not be reset and begins its count at the last converted point.

The comparator has two outputs. If the analog input is larger than the DAC output, the up output will be 1; and if the analog input is less than the DAC output, the down output will be 1. If V_A is higher than V_{DAC}, gate G_1 will be enabled, and the next clock pulse will set flip-flop 1, causing gate G_3 to have a 1 output and thereby enabling the count-up line of the counter. The counter will count upward until the DAC output is equal to V_A. The flip-flops are reset after

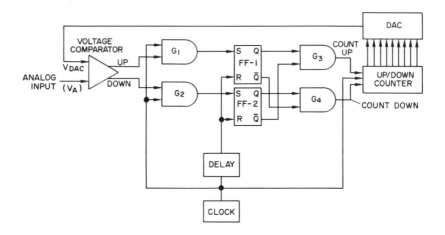

Fig. 10-14. The continuous A/D converter.

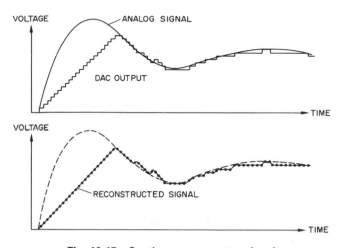

Fig. 10-15. Continuous converter signals.

each clock pulse by a delayed pulse from the clock (a one-shot can be used for this purpose). If V_A is less than V_{DAC}, the operation is the same, with the exception that the count-down line will be enabled, and the count will proceed in a downward direction.

The continuous ADC is able to follow an analog signal with a much closer reconstruction of the signal. This is shown in Fig. 10-15. The counter-type ADC, however, cannot follow a changing analog voltage as closely, as shown in Fig. 10-16.

MULTIPLEXING

Often it is desired to transmit a number of different analog signals simultaneously in a single digital channel. Further, often a single digital channel may output to a number of analog channels. For example, in a digitally controlled X-Y plotter, it may be desired to use only one DAC for both the X and Y channels. The process for doing this is called multiplexing.

Two methods of DAC are often used. For example, in Fig. 10-17 sequential digital data arriving from one source are distributed to a number of analog devices by switching the signal to separate DACs. The data is held in each analog channel, even when it is not being addressed by the storage register associated with each DAC.

A second method entails using one DAC and switching its analog output to separate analog devices through *sample-and-hold* amplifiers. This is shown in Fig. 10-18. The sample-and-hold amplifier is an op-amp with a capacitor at the input to store the DAC output level until the next sampling period. Although switches are shown, these are really solid-state switches. This second method is more economical, but it is necessary to renew the signal on the sample-and-hold

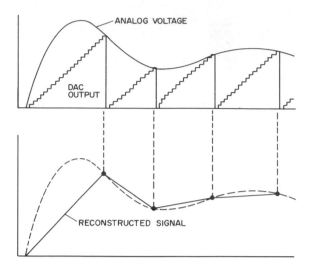

Fig. 10-16. Counter converter signals.

circuits at periodic intervals, while the preceding method stores the data indefinitely. This permits the automatic control of machines by precise positioning as, for example, in an automatic boring and milling machine.

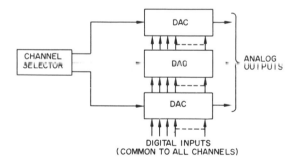

Fig. 10-17. DAC multiplexing using separate DACs.

ELECTROMECHANICAL ENCODERS

A very important application of ADC is in electromechanical devices that indicate angular and linear position, speed and direction of mechanical devices.

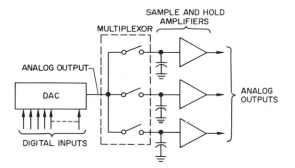

Fig. 10-18. Multiplexing using one DAC, multiplexer and sample-and-hold amplifiers.

Encoders can be either rotary shaft or linear straight-line types. A simple shaft encoder system is shown in Fig. 10-19. The encoder consists of a disc containing a pattern which may be sensed optically (light sensors), mechanically (brushes), or magnetically. The disc contains three tracks, 2^0, 2^1, and 2^3, to develop a 3-bit code indicating position. Note that the LSB equals $360°/2^3 = 45°$. If eight tracks were used, the degree of resolution would become $360°/2^8 = 360/256 = 1.4°$; a very high degree of resolution. As the disc rotates, it moves past one sensor per track to sense 0 and 1 and form the digital word.

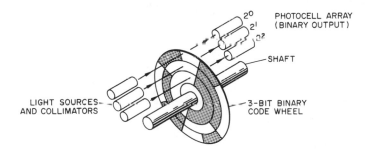

Fig. 10-19. A 3-bit rotary A/D encoder.

DIGITALLY CONTROLLED ANALOG DEVICES

Devices and machines that involve positioning in X and Y coordinates, and sometimes Z coordinates as well, are often controlled by a computer and hence

Digital Logic Circuits

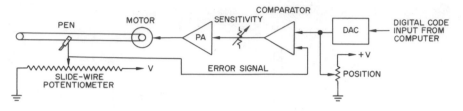

Fig. 10-20. Digitally controlled plotter.

digitally controlled. Examples are automated milling machines and X-Y plotters. The following describes digital control of an X-Y plotter.

An X-Y plotter consists of a pen which is moved over a sheet of paper in an X-Y coordinate format. The pen is positioned by two servo motors, one for the X axis and one for the Y. Figure 10-20 illustrates the circuit arrangement for one of the motor circuits; both X and Y are the same.

The computer generates a binary-coded word indicating a position along the axis. This data is fed to the DAC, converting it to an analog voltage which is fed to the comparator. The comparator compares the analog voltage to an error voltage proportional to the difference between the error voltage and the analog voltage. Further, the polarity of the comparator output will be positive or negative, depending on whether the analog voltage was greater or less than the error signal voltage. When both inputs to the comparator are the same, the comparator output will be zero. The comparator's output is amplified by a power amplifier (PA) which drives the motor in a clockwise or counterclockwise direction, depending on the polarity of the comparator output. When the pen has moved to a position causing the error voltage to equal the analog voltage, the motor will no longer be energized and the pen will stop.

Review Questions

1. If a code of 0110 is fed into a 4-bit binary scaled resistive divider network; what will the analog voltage output be if $V = +10$ volts?
2. In the circuit of question 1, what will be the output if the binary word is 1011?
3. In a 5-bit binary-scaled resistive divider, what will be the weights assigned to each bit? What will be the output for a digital word of 10110 if $V = 10$ volts?
4. If $V = 10$ volts in a 5-bit R-2R binary ladder network, what are the output voltages for each bit?
5. What is the advantage of the simultaneous-type ADC as compared to the counter-type?

6. Draw the output waveshape of the circuit shown below.

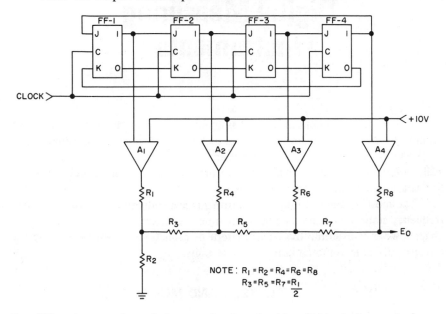

NOTE : $R_1 = R_2 = R_4 = R_6 = R_8$
$R_3 = R_5 = R_7 = \dfrac{R_1}{2}$

7. What degree of resolution can be found with a 7-bit shaft encoder?

Chapter 11
Digital Measuring Instruments

Today, direct numerical readout is making electronic measurements more accurate, faster, and easier. It reduces human error, since measurements are displayed as discrete numerals rather than a pointer deflection on a continuous scale. Digital instruments can also be interfaced to digital printers and computers for data logging and automatic analysis.

The most widely used digital instruments are voltmeters, multimeters, and frequency counters. Less frequent applications exist, such as L-C meters. In this chapter we will examine the basic system of each type of instrument and the circuitry of some commercially available units.

DIGITAL VOLTMETERS AND MULTIMETERS

Digital voltmeters and multimeters now permit a degree of measurement that was not previously possible. Low-cost meters typically provide an accuracy of 0.1 percent (dc), while expensive laboratory meters are as good as 0.001 percent. Although a number of techniques are in use, the most popular is the *dual-slope integrating* method. It measures the true average of the input voltage over a fixed measuring period accurately and in the presence of large amounts of superimposed noise.

A widely used digital multimeter is the Weston model 1240 shown in Fig. 11-1. It is compact and measures ac and dc voltage and current and resistance. It has ranges from 200 mvolts to 1,000 volts full-scale ($\pm 0.1\%$ accuracy), 200 µamps to 2 amps full-scale ($\pm 1\%$ accuracy), and 200 ohms to 20 megohms full-scale ($\pm 0.5\%$ to 1%).

The meter is basically an A/D converter with numeric readout, measuring either 200 mvolts or 2 volts dc. Alternating current (ac) and resistance are converted to direct current (dc) for measurement. The block diagram of the voltage-measuring ADC and its associated waveforms are shown in Figs. 11-2 and 11-3. The schematic diagrams are shown in Figs. 11-4 through 11-7.

Let us examine the ADC first (refer to Figs. 11-2, 11-3, and 11-4). From the volt/ohms input, a dc voltage is passed through a range-selector voltage-dividing network (Z_{101}) and R_{111} (limiting resistor) to the input of an op-amp integrator (pin 3, U_{101}). The range-selector circuit scales the voltage to either 200 mvolts or 2 volts. Transistors Q_{101} and Q_{102} are connected as zener diodes to suppress overvoltages.

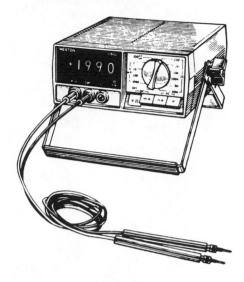

Fig. 11-1. The Weston model 1240 digital multimeter. (*Courtesy Weston*)

Op-amp U_{101} is an integrator which produces the ramp voltage waveform F (Fig. 11-3) from the charging of a capacitor (C in Fig. 11-2; C_{109} in Fig. 11-4). This integrating period is 200 msec (8,000 pulses from a 40-kHz clock), and is divided into ramp-up (t_1) and ramp-down (t_2) intervals. The circuit is driven from constant-current souces (Q_{104} positive ramp and Q_{107} negative ramp) to devlop linear ramps. During time t_1, the integrator charges to a voltage proportional to the input voltage. During t_2, the capacitor discharges an equal amount. The interval T (t_1 to t_2) is determined by the 8,000 clock pulses, and therefore the number of clock pulses during each interval varies with the voltage input. As the voltage increases, t_2 becomes longer and t_1 shorter.

The output of the integrator is fed to a comparator (U_{103}) which produces a 0 during t_1 and a 1 during t_2 (waveform G in Fig. 11-3). Waveform G is fed to the two inputs of the crossing shaper (pins 8 and 14 of U_{212}), one of the two signals being inverted by U_{210} (pin 8) to produce complementary inputs (refer to Fig. 11-5). The crossing shaper is a NOR gate producing a negative pulse (waveform H), which is fed to the reset inverter (U_{211C}) to produce a positive pulse to the set input of the ramp-down latch (U_{211C}), resetting it prior to the beginning at time t_2.

A 40-kHz clock oscillator (U_{210}) is divided down to 20 Hz (waveform A) by three decade counters (U_{201}, U_{202} and U_{203}) one divide-by-two (U_{213B}) counter. Then 10-Hz and 5-Hz signals are developed by counters U_{214A} and U_{214B} (waveforms B, C, and D). Three DCAs count up to 999 by IC's U_{201} through U_{209}. Nixie® tubes are used for readout.

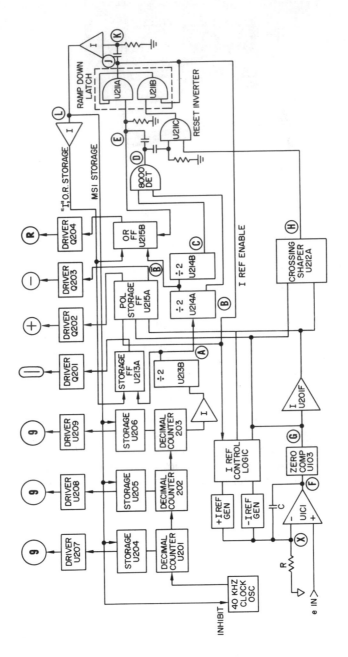

Fig. 11-2. Block diagram of Weston model 1240 meter.

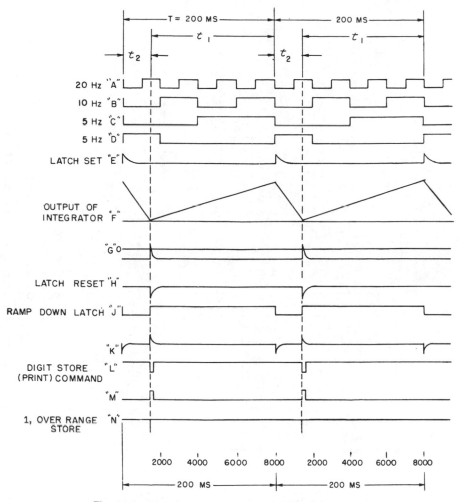

Fig. 11-3. Waveforms of Weston model 1240 meter.

The outputs of U_{214A} and U_{214B} have a 0 level on the 8,000 count (start of t_2) and are fed to the 8,000 detector, causing a 1 output (waveform D). The output of the 8,000 detector gate is differentiated by a capacitor (waveform E) and coupled to the ramp-down latch (U_{211A}) in order to provide an inhibit period of 6 μsec to prevent a race condition. The output of the ramp-down latch (waveform J) causes the storage registers associated with each readout to accept a new count from the decade counters. The count transferred will be the number of pulses developed during time t_2. During this time, the clock is momentarily stopped, preventing new pulses from entering the counter. At the end of waveform L's pulse, the registers latch at their new count and the readouts remain fixed in display.

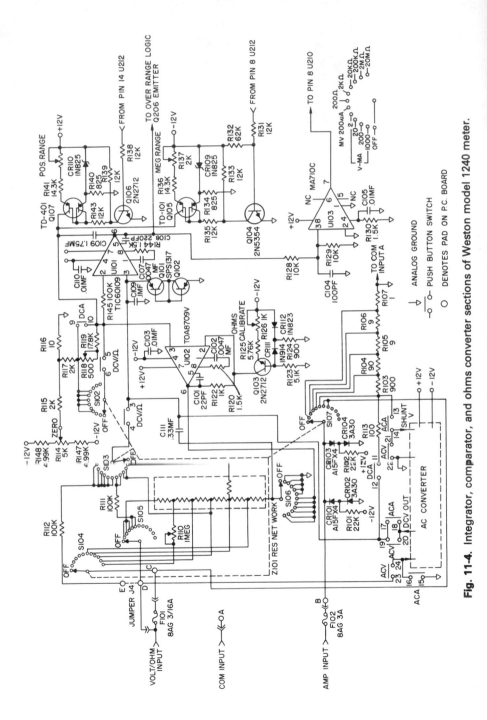

Fig. 11-4. Integrator, comparator, and ohms converter sections of Weston model 1240 meter.

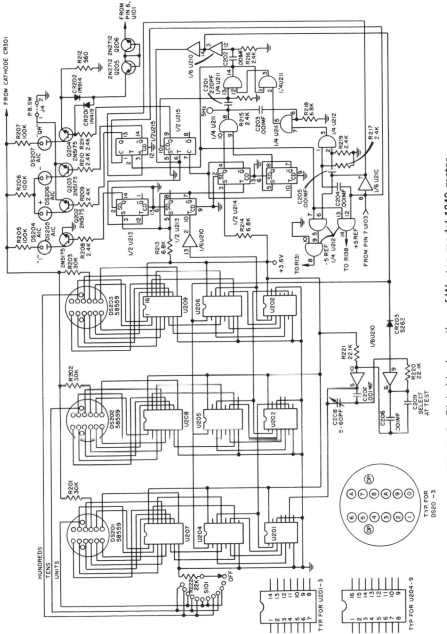

Fig. 11-5. Digital logic section of Weston model 1240 meter.

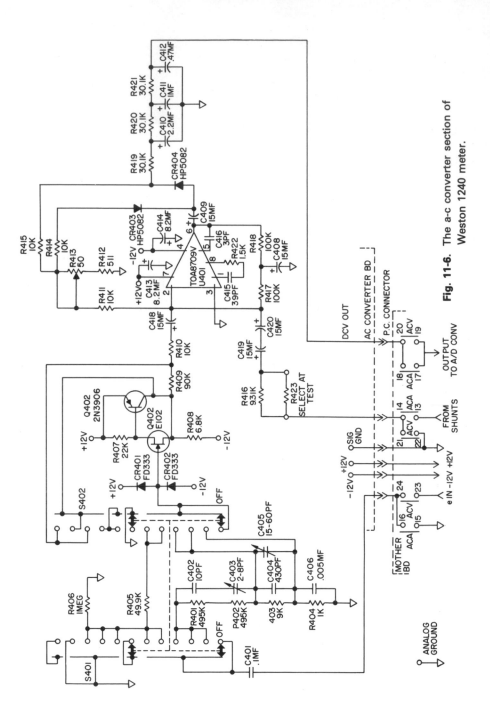

Fig. 11-6. The a-c converter section of Weston 1240 meter.

U_{213B} acts as a 1,000s counter, pulse L also causes transfer of the state of the counter into the storage register (U_{213A}) and display of the digit "1." Hence, the maximum count is 1,999. The over-range register (U_{215B}) and display (R) will be activated if a count of 2,000 is reached.

Resistance is converted to a dc voltage and applied to the ADC circuit previously described. IC U_{102} and transistor Q_{103} are a current source whose current is scaled by a voltage divider to provide 200 mvolts or 2 volts maximum input to the ADC. The voltage input becomes a function of the amount of resistance connected from the volt/ohms input to ground.

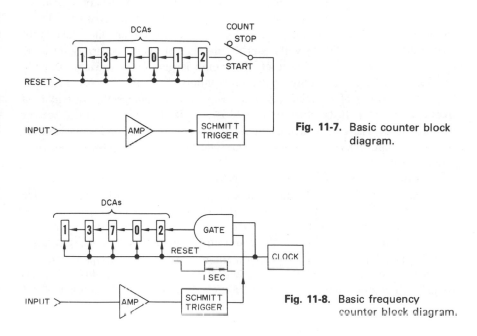

Fig. 11-7. Basic counter block diagram.

Fig. 11-8. Basic frequency counter block diagram.

Direct current is converted to voltage by passing the current through a series of precision shunt resistors. The shunts are protected from overload by diode clamps and a fuse. Alternating current is converted to d-c voltage by routing the voltage developed across the shunt through the a-c converter.

Alternating voltages are routed through the a-c converter (Fig. 11-6), which is essentially a half-wave average-sensing (RMS-calibrated) circuit. The input attenuator is frequency compensated, and its output is fed to a high-input impedance FET transistor operating as a source-follower/buffer to prevent the op-amp from being affected by the impedance variations of the attenuator. The op-amp has a gain of 1.11 to provide the conversion factor between average and RMS voltage. The output of the op-amp is rectified by the diodes CR_{403} and

CR_{404} and filtered by R_{420}, R_{421}, C_{410}, C_{411}, and C_{412} to develop the d-c input to the ADC.

ELECTRONIC COUNTERS

Digital counters have proven to be the most accurate, flexible, and convenient instruments available for making frequency and time-interval measurements. Counters are available for measuring frequencies from near zero to 40 GHz and time intervals from 100 psec to more than 100 days. They compare an unknown frequency or time interval to a known frequency or time interval.

The basic counting circuit is shown in Fig. 11-7. It consists of a series of DCAs (decade-counting assemblies) connected so that the carry of a DCA causes the next DCA to count. If the DCAs are reset to zero and then the control switch is closed, the DCAs will count the number of events occurring until the switch is opened or the DCAs reset. The amplifier and Schmitt trigger provide the necessary input sensitivity and signal shaping.

The addition of a clock and gate (Fig. 11-8) changes the system to a frequency counter. If the clock frequency is 0.5 Hz, its time period is 1 sec for the *on* period. The gate is then enabled for a period of 1 sec and will count the number of input pulses during the period. The DCAs are reset by the positive-going leading edge of the clock pulse.

A time interval, or period, is measured as shown in Fig. 11-9. The unknown signal produces a gating pulse, which causes the flip-flop to toggle twice per input cycle. This produces a gating pulse equal to the period of the input signal. During the gating time, the counter counts the 1-MHz clock pulses. At the end of the gating period, the count equals the period in microseconds.

A block diagram and a schematic diagram of a commercially available counter, Heath model IB-101, is shown in Figs. 11-10 and 11-11. It has frequency range of from 1 Hz to approximately 15 MHz, and an accuracy of ± 1 count. The input signal is amplified by a dual-gate MOSFET transistor (Q_1).

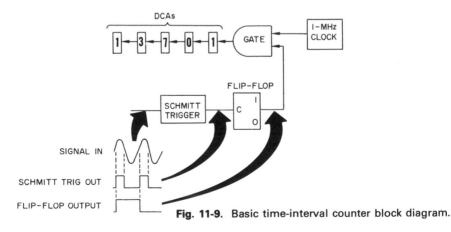

Fig. 11-9. Basic time-interval counter block diagram.

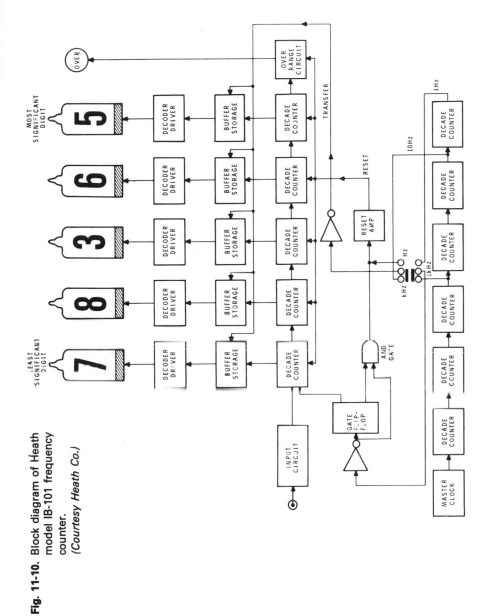

Fig. 11-10. Block diagram of Heath model IB-101 frequency counter. *(Courtesy Heath Co.)*

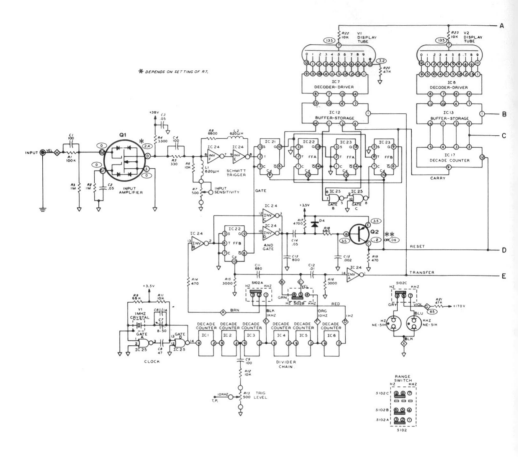

✱ *DEPENDS ON SETTING OF R7.*

ALL RESISTORS ARE 1/2 WATT UNLESS MARKED OTHERWISE.
   RESISTOR VALUES ARE IN OHMS (k=1000, M=1,000,000).

ALL RESISTORS ARE 5% UNLESS MARKED OTHERWISE.

ALL CAPACITOR VALUES LESS THAN 1 ARE IN µF. VALUES OF
   1 AND ABOVE ARE IN pF UNLESS MARKED OTHERWISE.

▽ THIS SYMBOL INDICATES CIRCUIT BOARD GROUND.

⏚ THIS SYMBOL INDICATES CHASSIS GROUND

◇ THIS SYMBOL INDICATES A LETTERED CIRCUIT BOARD
   CONNECTION.

⬭ THIS SYMBOL INDICATES A DC VOLTAGE TAKEN WITH
   A HIGH IMPEDANCE INPUT VOLTMETER FROM THE POINT
   INDICATED TO CHASSIS GROUND WITH NO INPUT SIGNAL
   TO THE COUNTER. VOLTAGES MAY VARY ±20%.

✱✱ WITH THE RANGE SWITCH IN THE HZ POSITION, THE
   VOLTAGE IS .06 VOLTS. WITH THE RANGE SWITCH IN THE
   KHZ POSITION, THE VOLTAGE IS .2 VOLTS.

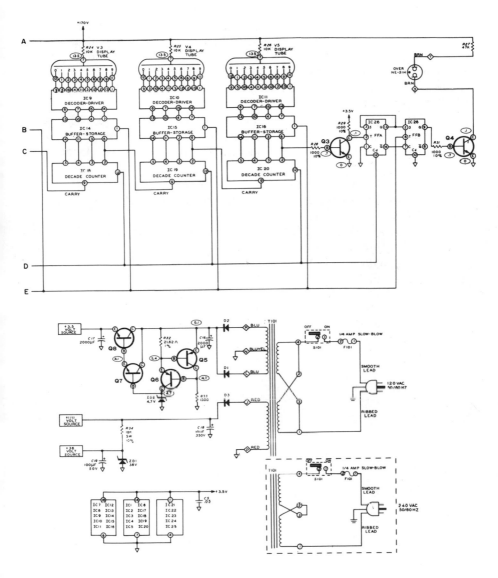

Fig. 11-11. Schematic of Heath model IB-101 frequency counter. (*Courtesy Heath Co.*)

Internal zener diodes protect the transistor against overvoltages. The output of Q_1 is converted to a pulse by IC-24 (E-F) operating as a Schmitt trigger. The output of the Schmitt trigger is fed to a DCA consisting of four J-K flip-flops and two NAND gates. The counter furnishes an 8421 code for the buffer/storage (IC-12), decoder/driver (IC-7), and display tube of the least significant digit. The DCA carry output activates the following DCA (IC-17), and so on for the remaining digits. The input signal will be counted until a timing signal is applied to the S and C inputs of FF-1 to inhibit the counter. A reset signal is applied to the Cd inputs of the flip-flops.

The buffer-storage stages accept a new count from the counters each time a transfer pulse is applied. the count is held and applied to the decoder/drivers. The appropriate cathodes of the display tubes are grounded by the decoder/drivers, causing digits to glow.

When a carry output is produced by IC-20 (pin 2), it is inverted (Q_3) and used to actuate the over-range lamp. IC-26 is the overload latch, and Q_4 is the lamp driver.

Gates A and D of IC-25 are a crystal-controlled multivibrator clock, to produce a square-wave 1-MHz signal. A divider chain, using six DCUs, divides the clock signal down to 1 kHz (IC-3 output) and 1 Hz (IC-6 output). When the range switch is in the *Hz* position the 1 Hz is fed to inverter B to toggle FF-B or

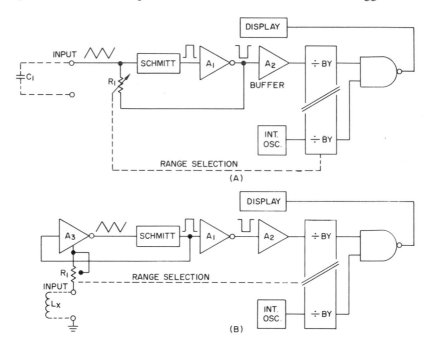

Fig. 11-12. Block diagram of the Systron-Donner model 9400 L/C digital meter. (*Courtesy Systron-Donner*)

IC-22. FF-B's output gates IC-21 for a period of 1 second, allowing the DCAs to count the incoming signal. During the time IC-21 is inhibited, the accumulated count is transferred, stored, displayed, and reset. In the *kHz* position of the range switch, a 1-msec (1 kHz) signal develops a gating enable period of 1 msec.

Inverters C and D of IC-24 form a NAND gate, producing a pulse at the same time IC-21 is inhibited. The pulse is differentiated by C_{12} and R_{16} and applied to inverter A of IC-24 to form the transfer pulse. In the *kHz* position, the transfer pulse (10 Hz) is taken from the output of IC-5.

The NAND gate's output is also differentiated by C_{14} and R_{17}, delayed slightly by R_{18} and C_{15}, and inverted by Q_2 to produce the reset signal.

OTHER DIGITAL INSTRUMENTATION

In addition to multimeters and frequency counters, the two most common digital instruments in use, a number of other instruments have gone digital. The application of digital logic to instrumentation is ever increasing. An example is the Systron-Donner model 9400 L/C meter, shown in Figs. 11-12 and 11-13. The digital L/C meter obsoletes the older, slower, and more error-prone comparison bridge. It permits precise measurements of capacitance from 0.1 pF to 100 μF and inductance from 0.1 μH to 100 mH. It also has a BCD (8421 code) output for connection to a digital printer or interfacing to a go/no-go comparator system such as is used on a production line for testing components.

As shown in Fig. 11-12(A), the capacitance mode utilizes a positive-feedback amplifier (Schmitt trigger), a d-c feedback amplifier, and the characteristics of an R-C timing circuit, to realize a constant-amplitude square-wave oscillator. The capacitor under test (C_1) charges at a rate of R × C until the positive trigger point of the Schmitt trigger is reached, causing it to switch states. This change of state is inverted by a unity-gain amplifier and fed back to the R-C time-constant circuit. Capacitor C_1 now discharges, again at a rate of R × C, until the negative trigger point of the Schmitt trigger is reached. This process will continue and provide an output pulse train which can be related to the values of R and C.

The output pulses are then buffered and divided down to be compared against a precise internal crystal-controlled oscillator. The divide-by circuits are arranged to scale the output pulse train into direct capacitance readings.

As the value of C_1 is changed, the frequency of the pulse train will change proportionately. Changing the value of R_1 (range) is accomplished by a front panel control to accommodate various values of capacitance.

The same basic procedure is followed in the inductance mode using the characteristic of an L/R time-constant. Of interest in this circuit is the ability to null out the resistance of the coil under test. This feature allows measurement of extremely low Q coils by simply adjusting the front panel control labeled *Series R Null.*

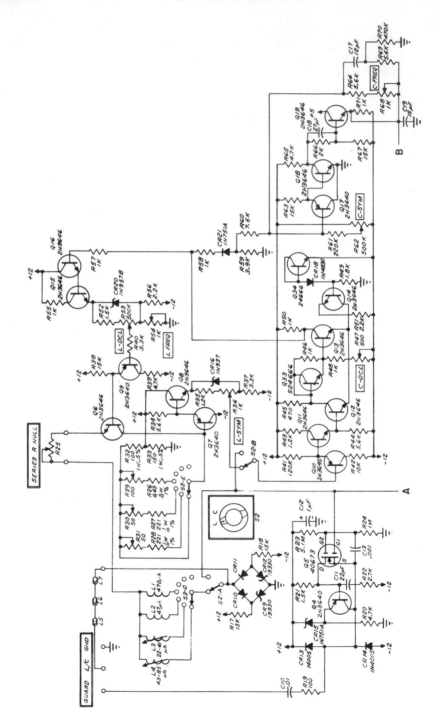

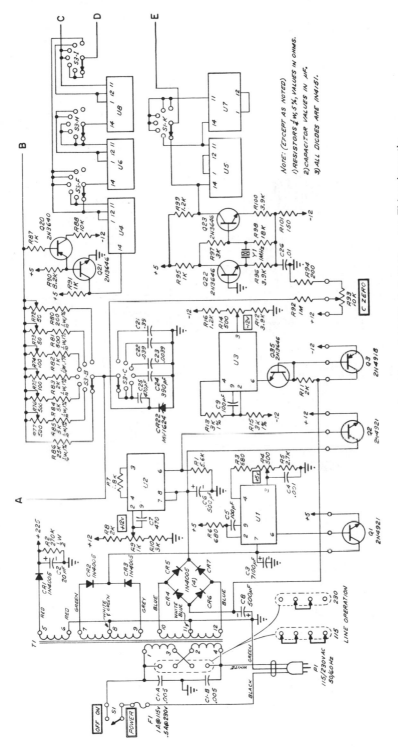

Fig. 11-13. Schematic of Systron-Donner model 9400 digital L/C meter. This schematic is continued on pp. 164 and 165.

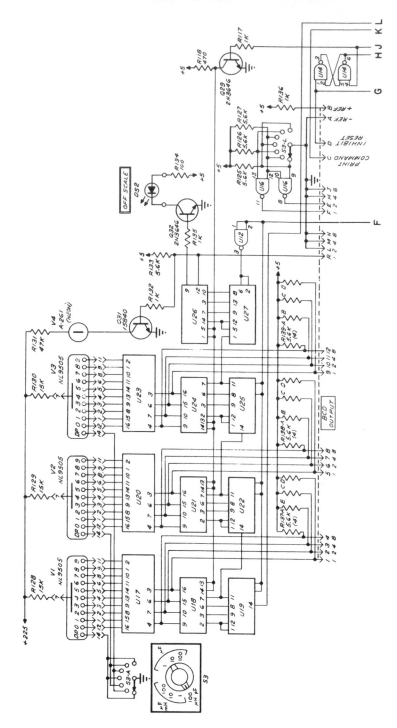

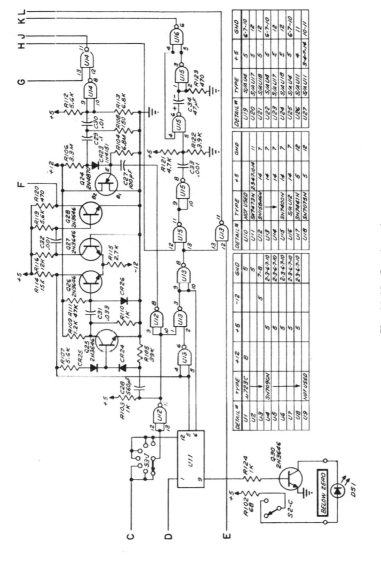

Fig. 11-13. Continued.

In the capacitance mode, switches S_{2A} and S_{2B} (Fig. 11-13) route the signal to Q_{10}, Q_{11}, and Q_{12}, serving as isolating and stabilizing amplifiers. Q_{33} and Q_{13} are the Schmitt trigger to convert the charging waveform to a pulse, which is amplified by Q_{17}, Q_{18}, Q_{19}, Q_{20}, and Q_{21}. ICs U_4, U_6, U_8, and U_{11} divide the frequency by the proper ratio for each range. The output is inverted by U_{12} and used to synchronize the firing of the unijunction transistor Q_{24}. The negative pulse produced at the emitter of Q_{24} is coupled through C_{30}, and inverted by U_{14C} to provide one input to NAND gate U_{14D}. The other input is provided by an inhibit reset line from the BCD output. When the inhibit reset line is at a 1 level, U_{14D} is enabled, and a positive pulse from U_{14C} generates a negative-going pulse from U_{14D}. Q_{29} inverts the pulse and drives the transfer line for the readout latching circuits (U_{18}, U_{21}, U_{24}, and U_{26}).

U_{12D}'s positive transient drives a one-shot (Q_{25} and Q_{26}) to set a minimum time (400 µsec) between gate openings. The leading edge creates the reset pulse through Q_{27} and Q_{28}, which sets the count chain to zero before each sample.

U_{12D} also drives one input of U_{13A}, while the other input is driven by a one-shot (Q_{25} and Q_{26}) through inverter U_{13B}. The sum of the two signals develops the gate time signal (U_{13B} output) which is inverted by U_{13C} to drive one input of U_{13D}. The other input comes from the clock or some division thereof. Summing these two signals provides a gated chain of pulses to drive the count chain (U_{19}, U_{22}, U_{25}, and U_{27}), which divides the signal by 10, thus providing a BCD output, while a single binary (U_{27A}) provides a full register of 1999. U_{27B} is a binary which changes states at count 2,000 to indicate over-range. ICs U_{18}, U_{21}, U_{24}, and U_{26} store the count between samples and provide the BCD output. U_{17}, U_{20}, and U_{23} are decoder/drivers for the display tubes. Q_{31} drives the "1" in the most significant digit display. Q_{32} drives the "off-scale" indicator.

The clock is a 1-MHz crystal-controlled oscillator (Q_{22}, Q_{23}, and Y_1) which is divided by U_5 and U_7 to provide the proper count frequency for each range. The inductance-measuring mode is similar to the capacitance-measuring mode with the measuring circuit consisting of an L/R oscillator.

Review Questions

1. What is the basic idea of a ramp-type DVM?
2. How many DCUs are employed in the Weston model 1240 DVM?
3. What is the maximum count on the Weston model 1240 DVM?
4. In the Weston DVM, what would be the trouble symptoms if IC-U210 failed?
5. What is the basic principle of a frequency counter?

6. In the Heath model IB-101 frequency counter, which IC functions as the gate?

7. In the Heath model IB-101, what would be the trouble symptoms if IC-6 failed?

8. In the Heath model IB-101, what is the function of IC-24E and F?

9. What is the basic principle of operation of the Systron-Donner L/C meter?

10. In the Systron-Donner L/C meter, how many DCAs and decoders are employed?

Chapter 12

Digital Calculators and Computers

Without doubt, the largest area of application of digital logic circuitry is in calculators and computers. The first production digital computer was developed by Remington Rand in 1951. By 1962, computers were fully transistorized, and 1970 saw the wide use of ICs. As component size and cost decreased, system speed and complexity increased. Further, the introduction of LSI (large-scale integration) and economical readouts made possible the introduction of low-cost, mass-produced digital calculators.

DIGITAL CALCULATORS

In the space of 2 or 3 years, the digital calculator has almost completely replaced the mechanical calculator. It is easier to use, faster, more accurate, smaller in size, and costs less. Calculators range from small battery-operated pocket units performing the basic four functions (addition, subtraction, multiplication, and division) to desk-top units that are programmable, interface with other units, and can do more than some computers.

The circuitry of a standard four-function, 10-digit calculator is shown in Figs. 12-1 and 12-2. The unit employs one 40-pin IC (type MK5010P) using

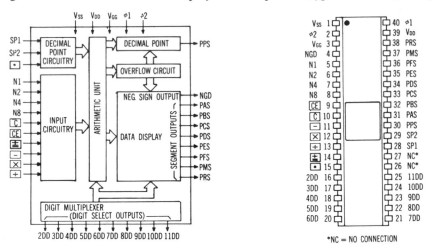

Fig. 12-1. MK5010P integrated circuit used in digital calculators: (A) functional circuit, and (B) pin connections. (*Courtesy MOSTEK Corp.*)

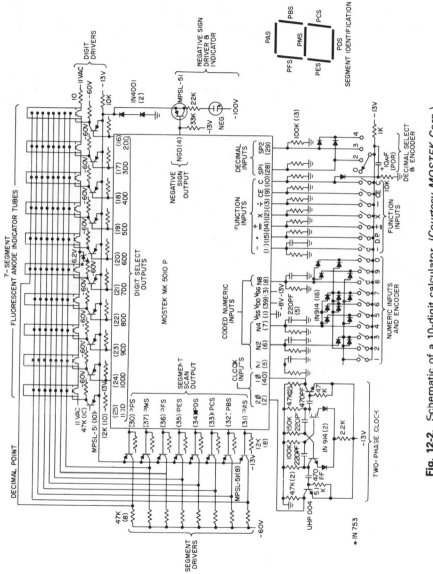

Fig. 12-2. Schematic of a 10-digit calculator. *(Courtesy MOSTEK Corp.)*

LSI MOS P-channel technology. The IC (Fig. 12-1) contains the digital logic. A free-running multivibrator (4-UHP004 transistors) provides the 18-kHz, two-phase clock signal to terminals 2 and 40 of the IC. Data is entered by depressing the digit keys and encoding to an 8421 code by the 16-diode encoding matrix (16-1N914 diodes; terminals 5, 6, 7, and 8 of IC). The inputs are sensed for their logic level; entry is made with 1 at any input, and previous entries are shifted to the left. With all inputs at 0, completion of entry occurs.

The numeric entry, or the resulting calculation, is displayed on a scanned basis. The seven segments are scanned in succession by IC outputs PAS through PPS (terminals 30–38). Output PPS is the decimal point output. A one-of-ten output at the digit select outputs enables the appropriate (selected) display digit. 11DD corresponds to the most significant digit (MSD), while 2DD corresponds to the least significant digit (LSD). As outputs are scanned from LSD to MSD, the appropriate data for each selected digit appears at the segment outputs. Segment data is ensured to be valid prior to enabling the digit select output.

By inhibiting the appropriate digit select outputs, the IC provides blanking of leading zeros; meaningless zeros to the left of the decimal point, except the 10^0 digit, are not displayed. All digits are blanked during calculation time.

Decimal point location may be selected to provide results in units only (0 places); in tenths (1 place); in hundredths (2 places); in thousandths (3 places); or ten-thousandths (4 places). This selection is made at inputs SP_1 and SP_2. Numeric entry may also be carried as far as the selected decimal point placement; further entry past this selected decimal point placement is disregarded.

COMPUTERS

The computer is essentially a calculator that can be programmed to input data automatically, perform operations, and output data. The sequence of instructions, called a *program*, may be changed easily by reading a new program into the machine. Such a unit is called a *general-purpose* computer. A *special-purpose* computer is designed and constructed to perform a fixed sequence of operation which cannot be changed except by changes in circuitry. This latter type of computer utilizes an ROM programmed for the specific task. We will concern ourselves here with the general-purpose machine which uses an RAM.

All general-purpose digital computers have the same fundamental structure, which consists of five basic sections. As shown in Fig. 12-3 (and Fig. 9-1), they are input, control, memory, arithmetic, and output. The design of these five basic sections varies significantly from one machine to another. Today there are several hundred different models on the market, with purchase prices ranging from a few thousand dollars to about ten million dollars! Most being manufactured are called *minicomputers*. These machines are physically small and relatively inexpensive. However, they are able to do many of the jobs done by

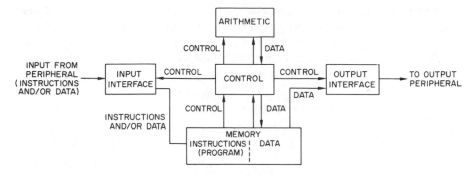

Fig. 12-3. Basic organization of a general-purpose computer.

large-scale computers, and can accommodate many of the peripherals used by medium- and large-scale systems.

It is difficult therefore to draw lines separating minicomputers, medium-, and large-scale machines. One method which can be used is word size. Minicomputers generally utilize 8- to 18-bit words. Medium-scale computers generally use 16- to 32-bit words. Large-scale machines generally use 24- to 36-bit words.

Most eight-bit word computers are really *controllers* used to control repetitive actions such as in industrial machinery. Their programs are usually not easily changed. The 12-bit word computers are used as controllers or minicomputers. They can perform single tasks, one task at a time, such as gathering data or performing analytical functions.

The 16-bit word computers are more complex and are capable of extensive multi-functional assignments. The 18-bit word machines can perform sophisticated analytical tasks and several functions simultaneously. The 24-bit word and larger machines are capable of highly sophisticated operations and performing many functions simultaneously. Keep in mind that there is a great deal of overlap in the ranges of these machines; for example, a large-size 12-bit word machine may be able to perform more sophisticated jobs than a minimum-size 18-bit word machine.

Computer Instruction Words

Recall that a word is a group of binary bits which may represent data (numeric or alpha characters). A word may also be used for instructions. Instruction words may be entered into a RAM in the same manner as data words. The instruction words may then be read from memory to the control section of the computer, causing a sequence of control signals and directing the arithmetic section to receive data, operate on that data, and output the data.

Generally, a single-address computer will start with an instruction word stored at a specific address in memory; it will read the word as an instruction and then continue reading instruction words from memory addresses, in order, until a *halt* or *branch* instruction is encountered. Instructions would be stored in one part of the memory while data would be stored in another part. For example, a computer capable of storing 4,096 words might store a program in addresses 0 to 1250 and store data in the remaining 2,846 addresses, or in a peripheral memory such as a magnetic tape unit.

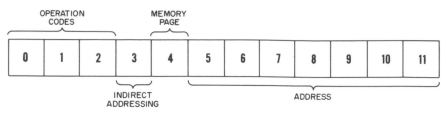

Fig. 12-4. Word organization used in PDP-8.

The memory, control circuitry, registers, and arithmetic sections are enclosed in one unit know as the CPU (central processor unit). A typical CPU is the PDP-8; its organization is shown in Fig. 12-5. The PDP-8 is manufactured by Digital Equipment Corporation and, as of this date, is the most widely used general-purpose minicomputer. The machine employs a single-address 12-bit word and 4,096-word memory (expandable to 32,768 words). It employs a 2's complement arithmetic unit, and multiplication and division are performed by *subroutines* (programmed sets of instructions). Input and output are most usually via a teletype terminal, although many peripheral I/O devices are available.

The PDP-8 instruction word is shown in Fig. 12-4. The first three bits (0 to 2) form the *op-code* (operation code), and there are hence eight specific instructions that can be stored in memory. Additional operations are accomplished by making one of the basic instructions a *jump-to-subroutine* instruction, by which the computer is directed to take its instructions from a particular set of addresses in memory. This greatly expands the instruction capabilities of a small op-code word length. The op-codes for the PDP-8 are shown in Table 12-1. A larger computer, having perhaps a five-bit op-code, could have a basic set of 32 instructions without resorting to subroutines which take up memory space.

The remaining nine bits (3 through 11) form the address portion of the instruction word. With only nine bits, we would normally only be able to address 512 locations. In order to address 4,096 memory locations, a technique is employed in which the memory is divided into 32 pages (blocks) of 128 words. The seven address bits (bits 5 through 11) of the instruction word can address any location in the page on which a current instruction is located, by placing a 1 in bit 4 of the instruction. A 0 in bit 4 allows any location in page 0 to be addressed directly from any page of memory. All other memory locations can be addressed indirectly by placing a 1 in bit 3 and a 7-bit address in bits 5 through 11, to specify the location of either the current page or page 0, containing the full 12-bit absolute address of the data. In other words, bit 3 determines if the addressing is direct (0) or indirect (1), and bit 4 determines if the address is in the current page (1) or first page (0) of the memory.

For example, if the instruction word is 0010 0001 0110, then the instruction (001) indicates that the operand quantity stored in memory at address location 22, page 0 (0 0001 0110), will be added to the quantity stored in the accumulator register. If the quantity at this location is 1,234 and the quantity in the accumulator register is 567, these two quantities will be added and the sum (2,801) will replace the quantity in the accumulator register.

Table 12-1. PDP-8 Computer Op-Codes.

Octal Code	Binary Code	Mnemonic Code	Description
0	000	AND-Y	The AND operation is performed between contents of memory location Y and contents of accumulator (AC). Result is left in AC (original contents lost), and contents of Y restored. May be considered bit-by-bit multiplication. Example: Original AC Y Final AC; 0 0 0; 0 1 0; 1 0 0; 1 1 1
1	001	ADD-Y	Contents of memory location Y added to contents of AC in 2's complement arithmetic. Result held in AC; original AC lost; contents of Y restored. If there is a carry, link is complemented (used in multiple precision multiplication).
2	010	ISZ-Y	Contents of memory location Y incremented by one in 2's complement arithmetic. If resultant contents of Y = 0, contents of program counter (PC) incremented by one and next instruction skipped. If resultant contents of Y – 0, program proceeds to next instruction. Incremented contents of Y restored to memory. Contents of AC not affected.
3	011	DCA-Y	Contents of AC put in memory at address Y and AC cleared. Previous contents of Y location lost.
4	100	JMS-Y	Contents of PC put in memory at address Y and next instruction taken from memory at address Y–1. Contents of AC not affected.
5	101	JMP-Y	Address Y set into PC so that next instruction is taken from memory address Y. Original contents of PC lost. AC not affected.
			Following two instructions not referenced to memory
6	110	IOP	Initiate operation (IOP) of peripheral equipment and information transfers between CPU and I/O device. Selection of equipment is designated by bits 3 through 8 of IOP instruction word.
7	111	OPR	Select either of two groups of 25 different microinstructions, to clear, complement, rotate, increment, etc.

If the instruction word is 0011 0001 0110, then the operand stored at address 1234 (address 82, page 9) will be added to the quantity in the accumulator register (567). If the quantity at address 1234 is 789, then these two quantities will be added, and the sum (1,356) will replace the quantity in the accumulator register.

CPU Organization

The basic organization of the PDP-8 CPU is shown in Fig. 12-5. Data in/out and communications between the various sections of the CPU is accomplished via a bus made up of 12 signal lines called the Omnibus®. The basic data paths within the PDP-8 CPU are shown in Fig. 12-6. The Omnibus is made up of the memory data (MD) and memory address (MA) buses. A third bus, the major register bus, also carries information but is not part of the omnibus.

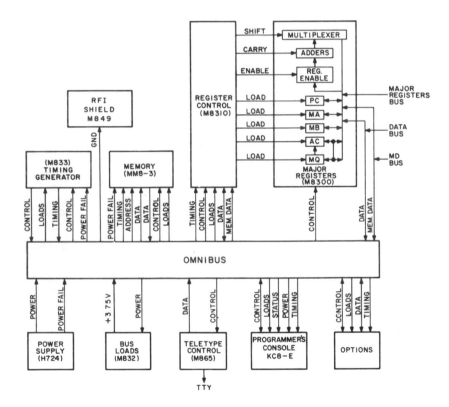

Fig. 12-5. Simplified block diagram of PDP-8/E CPU. Key: major registers: PC = program counter; MA = memory address; MB = memory buffer; AC = accumulator; MQ = multiply-quotient; TTY = input/output to/from teletype. (*Courtesy Digital Equipment Corp.*)

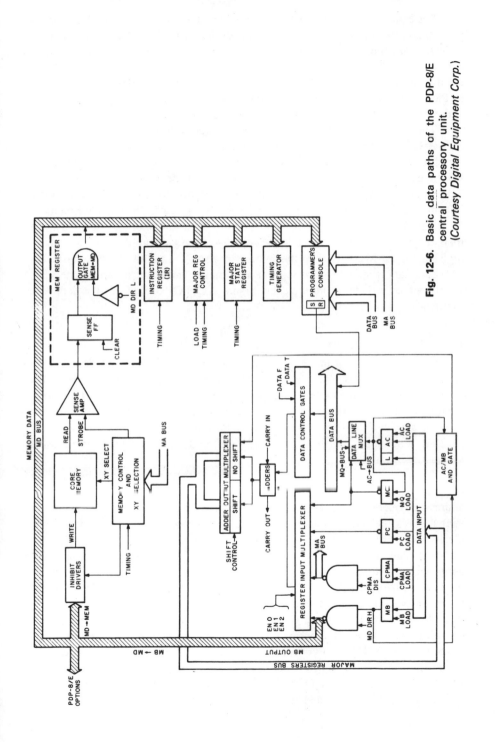

Fig. 12-6. Basic data paths of the PDP-8/E central processor unit. *(Courtesy Digital Equipment Corp.)*

For example, using the MD bus, memory data is provided by the memory register and the memory buffer (MB) register. The MA bus receives the memory address from the CPMA register and from some options that generate the 12 address bits. The MA bus applies these 12 bits to the X-Y selection decoder of the memory. The data bus is used to receive switch register data, provide status to the programmer's console, carry information to and from a peripheral unit, and provide a path for data to the major register bus.

The major registers bus completes a return path to each of the major registers. The CPU controls inputs to the major registers, and the enabling logic causes operations and loading to manipulate data and select one of the registers to store the results. If, for example, the results are to be stored into memory, the MB register is loaded and gated onto the MD bus by MD DIR. This same information is carried to the inhibit drivers and stored in the selected memory location. To place data stored in a memory location onto the data bus, the memory locations is first selected, the contents sensed and applied to the memory register. Signal MD DR gates the memory register out to the MD bus. From the MD bus, data is applied to the register multiplexer and to the major registers bus via the adder and output multiplexer. Signal AC-TO-BUS places the data onto the data bus.

CPU Memory System

The PDP-8 RAM memory is a coincident-current, magnetic core system similar to the one described in Chap. 8. The basic unit can store up to 4,096 twelve-bit words and can be expanded up to 32,768 words. The block diagram for the system, constructed on one plug-in printed circuit card, is shown in Fig. 12-7. The memory system decodes and selects the desired core location in which a 12-bit word is stored, or to be stored, and reads or writes the word into the selected location.

The operation cycle begins by loading the CPMA register and then placing the contents of the CPMA on the memory address (MA) lines. The memory address decoders receive the MA bits and turn on the corresponding read current switch when the return, source, and write are present. The memory register is cleared when the return and write become active.

The output from the 12 selected cores are fed to their respective sense amplifiers. A strobe signal is used to gate the sense amplifiers into the local memory register. If MD DIR is low (as it always is during the READ portion of the memory cycle), the output of the memory register is placed onto the memory data (MD) lines. During the write portion of the memory cycle, the memory selection system uses the same address inputs and control signals; however, control signal WRITE will change states, causing the write current switches to be activated. To write the contents of the memory register back into core, MD DIR will be low (active). Otherwise, the contents of the MB register will be placed on the MD lines, and the word in the MB register will be written into

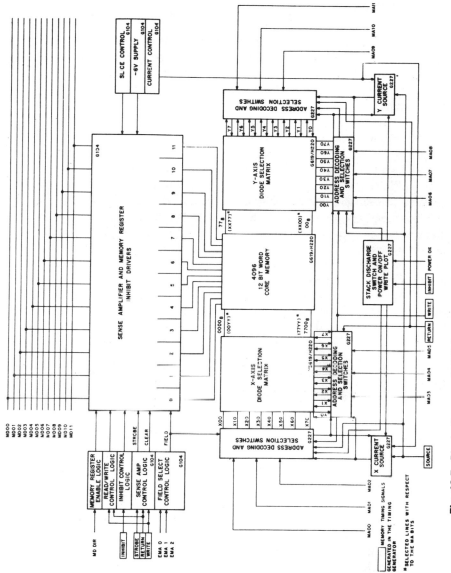

Fig. 12-7. Block diagram of PDP-8 memory system. *(Courtesy Digital Equipment Corp.)*

core. The inhibit signal controls the gating circuits, and only when INHIBIT is active will the inhibit drivers be activated. A 0 received from the MD lines and INHIBIT will cause the corresponding inhibit driver to produce inhibit current.

CPU Operation

The CPU manipulates data according to a program (a predetermined sequence of instructions). In the PDP-8, both the data to be manipulated and the instructions are stored in the memory. An instruction is brought from the memory to the control section of the CPU, where the instruction is decoded to determine what to do to the data and what data is affected. After the data has been manipulated, the result is stored within the control section, transferred to a memory location, or transferred to peripheral equipment. The CPU control section consists of the major register and register control modules, as shown in Fig. 12-8.

Data is transferred between registers, between registers and memory (via MD bus lines), between registers and peripherals (via data bus lines), and between register and internal options. Transfers are accomplished by a major register gating network which selects the particular register that takes part in the transfer and performs some operation on the data being transferred. The selection and the operation performed are determined by control signals developed within the major register control logic. This logic also provides control signals that determine the destination of the transferred data.

These signals that control the registers and their gating are developed within the major register control logic. The signals are developed largely in response to just three variables: (a) Basic instructions (as decoded by the instruction register logic), (b) processor major states (as determined by the major state register logic), and (c) time states and time pulses. These variables are combined to produce control signals that make the major register gating network function. For convenience, these signals are grouped according to function. Thus, source control signals select the register which contains the data that will be operated on; route control signals determine what is done to the data, and then place the result on the major register bus; finally, destination control signals load the result into the selected register.

Two other logic groupings are the skip logic and the link logic. Both these logic blocks utilize timing signals, major state signals, and instruction signals, among others, in carrying out operate microinstructions. The link, itself, is used to extend the arithmetic capability of the accumulator register. As shown, the link can be applied to the major register gating in response to shift signals and is also used to initiate the skip logic in response to certain operate microinstructions. The skip logic, in carrying out various microinstructions, is used primarily to generate the carry-in route control signals.

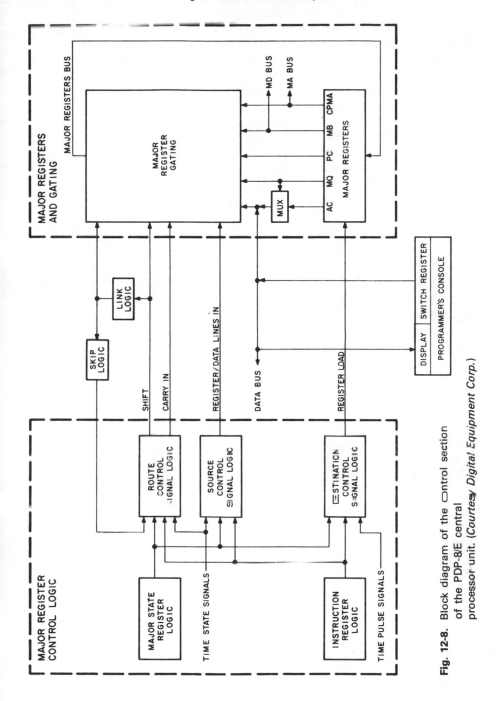

Fig. 12-8. Block diagram of the control section of the PDP-8/E central processor unit. (*Courtesy Digital Equipment Corp.*)

The programmer's console, although not physically a part of the CPU, is functionally inseparable. The operator can communicate with the major registers and cause data transfer to occur by operating various control panel keys. Data is transferred between the console and the CPU on the data lines, in response to control signals generated within the console logic shown in Fig. 12-8.

Review Questions

1. Write the truth table for the numeric input of the calculator shown in Fig. 12-2.
2. Referring to Fig. 12-2, if the numeral 1 is displayed in all digits, what will be the logic levels at terminals 30 through 37?
3. Write the PDP-8 instruction word (in binary form) that will add the operand quantity at address 121 of page 0 to the quantity in the accumulator register.
4. Write the PDP-8 instruction word (in binary form) that will transfer an operand quantity from the accumulator register to address 67 of the current page.

Index